PHILOSOPHY OF ECONOMICS

Edited by

WOLFGANG BALZER and BERT HAMMINGA

Reprinted from

Erkenntnis

Volume 30, Nos. 1–2, 1989

KLUWER ACADEMIC PUBLISHERS

DORDRECHT / BOSTON / LONDON

Library of Congress Cataloging-in-Publication Data

CIP

Philosophy of economics / edited by Wolfgang Balzer and Bert Hamminga.
p. cm.
Papers presented at a conference at Tilburg University, Netherlands, July 1987.
"Reprinted from Erkenntnis, volume 30, nos. 1-2, 1989."
ISBN 0-7923-0157-9
1. Economics--Philosophy--Congresses. I. Balzer, Wolfgang, 1947- II. Hamminga, Bert, 1951-
HB72.P47 1989
330'.01--dc19 89-31036

ISBN 0-7923-0157-9

Published by Kluwer Academic Publishers,
P.O. Box 17, 3300 AA Dordrecht, The Netherlands.

Kluwer Academic Publishers incorporates the publishing programmes of
D. Reidel, Martinus Nijhoff, Dr W. Junk and MTP Press.

Sold and distributed in the U.S.A. and Canada
by Kluwer Academic Publishers,
101 Philip Drive, Norwell, MA 02061, U.S.A.

In all other countries, sold and distributed
by Kluwer Academic Publishers Group,
P.O. Box 322, 3300 AH Dordrecht, The Netherlands.

02-091-150 ts

Printed in the Netherlands

ERKENNTNIS / *Volume 30 Nos. 1–2 March 1989*

PHILOSOPHY OF ECONOMICS

Guest Editors

WOLFGANG BALZER and BERT HAMMINGA

INTRODUCTION

The last decade witnessed an unprecedented annual growth of the literature dealing with the philosophy of economics, as well as the first signs of an institutionalization (conferences, an international journal) of the philosophy of economics as a scientific subject in itself – in particular in the U.S. In 1981 a meeting took place with participants mainly of European "continental" origin. In July 1987, we organized a second conference "Philosophy of Economics II" at Tilburg University, The Netherlands, mainly aiming at the establishment of first contacts between the middle-European group and researchers from the U.S. The present volume contains the papers presented at this conference.

Philosophical thought on economics in recent years split up in many different streams, two of which are represented in the larger part of this volume.

The first of these streams was formed by a group of researchers mainly from middle-Europe, who make empirical studies of the logical structures of the different theories as they find them presented in economic literature. Two methods prevail here. First, the structuralist method, as exemplified in the writings of Sneed, Stegmüller and others, of describing the object of a theory as a set of ("partial potential") models. Such models consist of sets and relationships between these sets, which represent the concepts used in the theory. The method leads to a precise study of the way in which out of the entire class of possible models, the axioms of the theory select those models with which the axioms are consistent. The method involves, in itself, hardly any philosophical a priori: it is applicable to any assertion made with the help of concepts. The natural perspective in applying the method is a classification of such assertions (among which presentations of scientific theories are of special interest) according to their model-theoretic skeleton, analogous to the classification of chemicals on the basis of their molecular configurations which was made possible by the method of structural analysis developed in chemistry in the nineteenth century.

Erkenntnis **30** (1989) 1–3.

A second approach in the same direction was developed by the Polish scholar Leszek Nowak. His method proceeds from the principle that economic theories, and, indeed, scientific theories in general, do not yield a completely correct description of their objects, nor are they constructed for that purpose. Instead, scientists believe that their laws would hold under certain idealizing conditions. This belief is not testable directly, since the idealizing conditions are never completely satisfied. From this viewpoint, reconstruction of a theory means, first, to find the relevant idealizing conditions, second, to study the way in which scientists "concretize" their laws by successively dropping their idealizing conditions, and third, to study the relationship between the ideal, nonexisting objects for which the scientist believes his theory would hold, and the real objects that the theory is ultimately intended to describe.

The second stream, consisting largely of Anglo-Saxon philosophers of economics, is characterized by a less formal style, a different way of choosing problems and of writing about them. Their discussions arose mainly from the normative and epistemological puzzles which one has to face when one tries to apply to economics any criteria for "good" (as opposed to "pseudo") science developed by philosophers of science such as Karl Popper and Irme Lakatos. It seems that many of these criteria plead for a type of economics that does not occur in the real world of working economists, while the types of economics actually practised are in danger of coming out as pseudo-science. In the Anglo-Saxon discussion the puzzles raised by these criteria have lost some prominence, but the priority of normative-epistemological problems remains: Can economics be called a science? To what extent? What is characteristic of an economic argument? What predictive power should an economic theory have? How can micro-economic theories justify macro-economic theories?

The papers delivered at the Tilburg conference have been arranged here according to this main distinction. The first group contains contributions to epistemological and normative issues (Hausman, Nelson, Rosenberg). It is followed by some intermediary articles (Bicchieri, de la Sienra, Pearce and Pearce). The third group is of a clearly structuralist flavour (Balzer and Haendler, Diederich, Janssen, Kuipers and Janssen, Sneed), and the final group (Nowak, Hamminga), deals with the problem of idealization.

The conference at Tilburg University was financially and materially supported by the Dutch *Organisatie voor Zuiver Wetenschappelijk Onderzoek* (ZWO), and by the *Cobbenhagenfonds* and the Department of Philosophy, both of Tilburg University.

We are grateful to all those who have, in different ways, contributed to the successful organisation of the conference, of whom we want to thank especially Prof. Dr. M. A. D. Plattel and Mr. M. A. M. van den Akker. We are also indebted to the editors of *Erkenntnis* for their decision to publish this collection as an extra double issue of the journal.

Universität München
Seminar für Philosophie, Logik
und Wissenschaftstheorie
Ludwigstr. 31, 8000 München 22
F.D.R.

WOLFGANG BALZER
BERT HAMMINGA

DANIEL M. HAUSMAN

ARBITRAGE ARGUMENTS

Consider the following famous argument, which motives accepting the hypothesis of rational expectations:[1]

(RE: The rational expectations argument) I should like to suggest that expectations, since they are informed predictions of future events, are essentially the same as the predictions of the relevant economic theory.... If the prediction of the theory were substantially better than the expectations of the firms, then there would be opportunities for the "insider" to profit from the knowledge – by inventory speculation if possible, by operating a firm, or by selling a price forecasting service to the firms. The profit opportunities would no longer exist if the aggregate expectation of the firms is the same as the prediction of the theory:... (Muth 1961)

This is a powerful argument, although some reasonable doubts may remain. The phrase "the predictions of economic theory" is elliptical in an important way, for economic theory makes predictions only with the help of information concerning initial conditions, which may not be shared by various economic agents. Furthermore, it might be the case that economists cannot make a killing not because the expectations of firms match the predictions of economic theory, but because the expectations of firms are superior to economic theory. One is also reminded of Keynes' strikingly successful efforts on the stock market on behalf of King's College. But clearly there is something to the argument – enough to convince many talented economists to take seriously the rational expectations hypothesis.

(RE) is a simple instance of a general form of argument that is much used and sometimes abused not only by economists but by others as well. Arguments of this form I shall call arbitrage arguments because they turn crucially on the possibility of exploiting disequilibrium exemplified by arbitrage. What is the form of arbitrage arguments, and when is it reasonable to be persuaded by them?

Let us first reformulate (RE) as a logically valid argument:

(RE′)

(1) (existential premise) The expectations of some firms match the predictions of economic theory.

Erkenntnis **30** (1989) 5–22.

(2) (fitness premise) If the expectations of firms that match the predictions of economic theory are in some domain more accurate than expectations of other firms, then these firms will, ceteris paribus, make larger profits than firms whose expectations do not match the predictions of economic theory.

(3) (equilibrium premise) It is not, for the most part, the case that firms run by economists or that employ relatively many economic consultants make larger profits than other firms.

(4) (adaptation premise) The predictions of economic theory are ceteris paribus more accurate than expectations generated in any other known way.

(5) (indicator premise) Firms run by economists or that employ relatively many economists as consultants have expectations that match the predictions of economic theory.

Therefore

(6) (conclusion) Ceteris paribus, the expectations of firms for the most part match the predictions of economic theory.

As stated this argument would not satisfy a logician's account of validity.[2] The interpretation of ceteris paribus clauses is a tricky matter and the vague quantification of "for the most part" needs clarification. But it seems to me that provided that the ceteris paribus clause in the conclusion encompasses the ceteris paribus clauses in the premises, one is stretching only a little in regarding RE′ as a strictly valid argument – that is, as an argument in which it is logically impossible for all the premises to be true and the conclusion to be false. It is helpful to recast an argument so that it is logically valid. For in that case, all questions about its *soundness* can focus on the truth of the premises.

In arbitrage arguments typically authors state merely that some firms *could* possess the property that would give them a competitive advantage, but without the premise that some firm actually *does* possess that property (which in this case is possessing expectations that more or less coincide with the predictions of economic theory), the argument would not be logically valid. One might instead formulate arbitrage arguments as counterfactuals with premises concerning what could happen and conclusions concerning what would; but it is better to avoid the ambiguities and unclarities that counterfactuals introduce. Thus the somewhat strange formulation of premise (1), which I have called the existential premise.

The names for (2) and (4) are borrowed from biology, where they are used in the sense employed here. It may seem a waste of time to saddle (2), the fitness premise with a long antecedent that is then asserted in the adaptation premise (4), but, as we shall shortly see, arbitrage arguments can be used to argue against analogues to the antecedent of the fitness premise (2) just as easily as they can be used to argue that all agents must possess some property or other. Note the ceteris paribus clause. If, for example, it turned out that firms whose expectations match the predictions of economic theory were generally led by timid and muddle-headed individuals, then it might not be the case that these firms would earn higher profits. The ceteris paribus conditions in (2) and its analogues in other arbitrage arguments are often ill-specified and hard to satisfy.

The equilibrium premise, (3) is in principle a straight-forward empirical claim. Note that no one is asserting that *no* firm run by an economist makes extremely high profits, just that on average firms obviously led by the wisdom of economic theory do not do better than firms not so led.

(4), the adaptation premise, simply makes explicit the presumed superiority of economic theory. It is, as we shall see, often the weakest link in an arbitrage argument in economics. There might be some question about whether the ceteris paribus clause in (4) is necessary, given the usual stochastic construal of "accuracy" in rational expectations equilibrium. But it might be the case that somebody's predictions are more accurate than those derived from economic theory not because of luck, but because he or she has actually discovered a better theory. It is, in any event, safer to regard the clause as present. In other arbitrage arguments the ceteris paribus clause in the adaptation premise will be more significant. (Note that the ceteris paribus clause in (6) is needed regardless of whether one regards (4) as carrying such an implicit qualification.) (5) is an indicator premise that operationalizes the idea that those whose expectations match the predictions of economic theory will make larger profits than those whose expectations do not by telling us whose expectations match the predictions of economic theory.

The conclusion to the argument, finally, inherits both the ceteris paribus clauses of the fitness and adaptation premises and the "for the most part" clause of the equilibrium premise, (3). All one can validly conclude from these premises is that, other things being equal, the expectations of firms will be distributed around the predictions of economic theory with whatever systematic differences permitted by the degree of approximation in the equilibrium premise.

Is (RE′) sound – that is, are premises (1)–(5) true? (1), the existential premise seems acceptable in contemporary applications of the hypothesis of rational expectations. Surely it is the case, for good or ill, that some firms take the relevant economic theories seriously and are guided by them. If in applications of the hypothesis of rational expectations to the past, however, it is supposed that the expectations of firms matched the predictions of economic theories which were not yet known, then the existential premise might be dubious. Appearances to the contrary, the fitness premise is far from obvious. If one's knowledge of the future is extremely weak and defective, improvements will not necessarily lead to competitive advantage. One might, for example, find an analogue here to the theory of the second best. Further questions might be raised concerning the extent of the fitness, given the extent of the superiority of knowledge of the future.

The equilibrium premise, (3) seems to be a straight-forward and reasonably well-supported empirical claim. There are, however, empirical complications involved in its confirmation: if the fitness conferred by the wisdom of economic theory is fairly small, then it might be impossible, given the noise in the data, to detect the larger profits firms with such wise expectations make.

There is obviously some dispute about the truth of (4), the adaptation premise – that is about the predictive value of economic theory – although most orthodox economists seem convinced of the predictive value of orthodox theory (and similarly for non-orthodox theorists and their own theories).[3] As we shall see shortly, it is also possible to make a rather precarious arbitrage argument for the truth of this premise. Finally, although one might quibble about the indicator premise (5), it clearly seems reasonable to take firms that are evidently guided by those schooled in economic theory as firms whose expectations match the predictions of economic theory. In considering the argument one must also consider the ceteris paribus clause in the conclusion, for Muth and others downplay this important qualification.

So (RE) is neither absurd nor obviously sound. It may reasonably persuade people to take its conclusion seriously and to seek further evidence. To account for the persuasiveness of Muth's argument, it is thus not necessary to follow McCloskey (1985, Chap. 6) in stressing the role of rhetorical devices such as analogy. On the other hand, the existence of such a valid and arguably sound argument does not imply that there are no extra-logical means of persuasion at work here. There is no reason why any argument may not *both* stress an analogy and be logically sound.

Not all arbitrage arguments have just this structure. To show some of the possible complexities, consider the following famous argument in population genetics, originally due to R. A. Fisher (1931, pp. 158f). Here is Hamilton's formulation:

(1 : 1 Sex ratio argument [my name])

(1) Suppose male births are less common than female.

(2) A newborn male then has better mating prospects than a newborn female, and therefore can expect to have more offspring.

(3) Therefore parents genetically disposed to produce males tend to have more than average numbers of grandchildren born to them.

(4) Therefore the genes for male-producing tendencies spread, and male births before commoner.

(5) As the 1 : 1 sex ratio is approached, the advantage associated with producing males dies away.

(6) The same reasoning holds if females are substituted for males throughout. Therefore 1 : 1 is the equilibrium ratio. (Hamilton 1967, p. 477)

1 : 1 differs from RE in two main ways. First, it is concerned with the equilibrium sex ratio in the population, not necessarily with the equilibrium sex ratio among offspring of particular members of the population. In equilibrium, if there are as many organisms producing offspring in the ratio N : M as there are producing in the ratio M : N, then there will be no selection pressure against either of these ratios. In contrast the RE argument is concerned with the behavior of the individual members of the population of firms. It supports a more refined conclusion. Second, different behavior is adapted to circumstances of disequilibrium than in circumstances of equilibrium, while

in RE theories optimal behavior often varies little, and, in any case, disequilibrium scarcely obtains.

Let us then reformulate Hamilton's argument so that it roughly parallels RE′ and then use this reformulation to suggest a general schema for arbitrage arguments.

(1 : 1′)

(1) (existential premise) In any population, some pair of organisms will produce male and female offspring in whatever ratio is best adapted to the particular environment in which the population finds itself.

(2) (fitness premise) If for some N, M, in some environment, a pair of organisms that is genetically disposed to produce male and female offspring in the ratio $N : M$ has ceteris paribus more grandchildren than average, then ceteris paribus the genes for producing offspring in that ratio will spread.

(3) (equilibrium premise) It is not for the most part the case that genes for producing offspring in the ratio $N : M$ are spreading.

(4) (adaptation premise) If the ratio of males to females in the population is larger (smaller) than 1 : 1 and $N < M$ ($N > M$), then a pair of organisms that tends to produce male and female offspring in the ratio $N : M$ has in that environment, ceteris paribus more grandchildren than average.

(5) (indicator premise) Pairs of organisms that produce male and female offspring in the ratio $N : M$ are genetically disposed to produce male and female offspring in that ratio.

Therefore,

(6) Ceteris paribus the ratio of male to females is close to 1 : 1.

Apart from the slack of the ceteris paribus clauses and the "for the most part" of (3) and the "close to" of (6), this is a logically valid argument. Given the negation of (6), and (1), (2), (4) and (5), it follows that (3) must be false. Hence one can validly infer (6) from (1–5). (5) is, of course, badly oversimplified, although (I hope) in a harmless way.

The differences between the 1 : 1 and RE arguments show up in the adaptation premise and in the less specific conclusion. Although there may well be some forms of arbitrage arguments that have a more complicated structure, let me now venture the following first stab at a general form for such arguments.

Arbitrage Argument Schema:

(i) (existential premise) For any feasible environment e and any property H in a set of properties $\{H(e)\}$, some individuals x (in population P) possess property H.

(ii) (fitness premise) For any environment e, property H and individual x, if the possession of H is adaptive (leads to possession of property A_e) and not all individuals possess H, then, ceteris paribus, the possession of H confers a relative fitness on x (x does better than average according to some criterion C).

(iii) (equilibrium premise) It is not in general the case in the actual environment e^* that, ceteris paribus, any individual is relatively fit – does better than average according to some criterion C.

(iv–A) (adaptation premise) For all x, in the actual environment e^* the possession of property H^* is adaptive (leads ceteris paribus to possession of property A_{e^*}).

or

(iv–B) (diversity premise) Some x in the actual environment e^* do not possess H^*.

Therefore

(v–A) (conclusion A) Ceteris paribus, in the actual environment almost all x possess H^*.

or

(v-B) (conclusion B) It is not the case that ceteris paribus for the most part in the actual environment e^* that possessing H^* is adaptive.[4]

Since (as in the 1 : 1 argument) different properties may be adapted to different environments, we need to consider a set of properties $\{H(e)\}$ and various environments e including the actual environment e^*. In some cases, as in the RE argument, we may be able to draw strong conclusions (ceteris paribus) concerning properties of almost all members of the population. In other cases, as in the 1 : 1 argument, we may only be able to draw a weaker conclusion concerning some property of the population in the actual environment, but we can treat properties of populations as properties of the individuals that make them up. I will discuss and illustrate arguments employing (iv–B) and arriving at (v–B) shortly.

In the case of the argument for rational expectations, the population consists of firms, the properties H consist of the single property (which is also H^*) of having expectations that match the predictions of economic theory. The adaptive property A_e (which, like fitness is not a member of $\{H(e)\}$) is the property of possessing (relatively) accurate knowledge of the future; and criterion C is the rate of return on investment. Given the additional premise that to possess expectations that match the predictions of economic theory is to possess the most accurate expectations, the conclusion that the expectations of all firms match the predictions of economic theory follows. The environment is left implicit, but obviously the argument assumes a competitive economy. I left the indicator premise out of the general schema in order to focus on the essential elements.

In the case of the sex ratio argument, the population consists of some population or organisms that reproduce sexually. The properties He, x consist of the differing genetic tendencies to produce male and female offspring in various ratios. In addition to general features of the environment, e depends on the actual sex ratio in the population. H^* is not the property of producing offspring in a 1 : 1 sex ratio, but the different property of belonging to a population that produces offspring in a 1 : 1 sex ratio. A_e is the property of producing more grandchildren than average. The criterion of success, C, is gene propagation.

The general form provides one with a choice. The existential, fitness and equilibrium premises (which are the core of the argument) jointly imply *either* that the population possesses some property (or that all members of the population do) or that possessing H^* does not make an agent adapted to the actual environment – that is, possessing H^* does not lead it to possess A_{e^*}. To conclude that one of the disjuncts is true, the other must be denied, which is accomplished either by the adaptation or the diversity premises. Thus a cynic might prefer to insist on the diversity premise (which is the denial of the conclusion of the argument for the rational expectations hypothesis) and arrive instead at the conclusion that the predictions of economic theory are not more accurate than expectations generated in other ways.

The basic structure of arbitrage arguments should by now be transparent. If there is an equilibrium and a certain trait, which is supposedly advantageous, is found in some members of the population, then either it must be found in all members (or in population features) or it must not in reality be well adapted after all. Such arguments can be made in any domain in which the notions of fitness and of an equilibrium that could be shifted by a comparative advantage apply. Economics and population genetics are the obvious domains, but there are potential applications to other fields as well, such as anthropology or even chemistry.[5]

Although commentators on economics are often leery of the postulate of equilibrium, it seems to me that the equilibrium premise (within some suitable degree of approximation) is generally fairly robust and unproblematic. So, I suggest, in the case of economics are the existential and diversity premises. The premises to be skeptical of in economic applications are particularly the adaptation and fitness premises, which are either dubious in themselves or problematic on account of their ceteris paribus clauses. In evolutionary biology there are also special difficulties concerning the existential premise (Elster 1979, Chap. 1), for mutation may never stumble upon adaptations that are obvious to human foresight and intelligence.

As Elliott Sober has noted (1983) equilibrium explanations – arbitrage arguments that are employed to explain their conclusions – possess a special power and attraction. Without specifying the causal mechanism or the causal initial conditions that are responsible for the particular state of the population or of its members, an arbitrage argument can show that state to have been more or less inevitable. As a good explanation should, such an argument eliminates mere contingency and shows that what is in fact the case is what one would have expected, if one had known the laws and the relevant initial conditions. Equilibrium explanations are particularly powerful explanations because so little depends on the precise initial conditions.

But the price of this power is a peculiar precariousness. In a more specific causal explanation, one identifies the particular causes or causal conditions upon which the explanandum depends and the law or laws which link cause and effect. Such explanations can easily be incorrect, of course. We may be mistaken in our purported knowledge of the purported laws. We may be mistaken in our claims concerning which causes or causal conditions were present. We may mistakenly believe that the ceteris paribus qualifications, which attach to all causal claims are met. But since the explanation is based on knowledge of the particular causal circumstances, there are limits to such failings. In equilibrium explanations, in contrast, one does not have to identify the causes and the relevant laws; the explanandum is supposed to obtain, ceteris paribus regardless of the particular causal antecedents. But without knowing the particular causal history, one is often in a poor position to check claims of adaptation or to know whether the ceteris paribus condition is actually met. In the case of the rational expectations argument, for example, no one checks to see whether particular firms in a particular (disequilibrium) environment that employed neoclassical economic theory actually made better forecasts than did firms that did not employ neoclassical theory. No one checks to see whether in those circumstances the better forecasts actually led those firms to do better. No one does much more than guess whether there were no other "interferences" or "disturbing causes" that led firms employing economic theory to do worse or to make poorer forecasts, despite the predictive virtues of the theory.

Let me illustrate the power and the pitfalls of arbitrage arguments with two rather different examples from philosophy and economics, which are instances of the general schema and show weaknesses at exactly the suggested spots. The first is from a philosophy paper and is an arbitrage argument for the accuracy of the adaptation premise (4) in (RE):

(PS: Predictive Success Argument) Contrary to Rosenberg's claim, predictive success *is* an important criterion of theory choice in economics. One of the reasons for this is that economic predictions are *consumed* by the business community. . . . The argument is only that the survivability of the traditional approach in such applications indicates that (relative to the available alternatives), its predictive failings are not as great as Rosenberg would have us believe. (Hands 1984, p. 498)

Recasting this argument so that it is (with some differences to be discussed) an instance of the general schema for arbitrage arguments helps one to see clearly where it breaks down and what the general pitfalls of the argumentative strategy are. I suggest the following reformulation:

(PS′)

(1) (existential premise) Some firms do not pay for the predictions of economic theory and instead employ a cheaper alternative means of making predictions.

(2) (fitness premise) If firms that are not paying for predictions of economic theory are paying less for information that is just as accurate as economic theory, then, ceteris paribus, those firms will have higher net revenue.

(3) (equilibrium premise) In general, firms that do not pay for the predictions of economic theory and employ some alternative do not have higher net revenues.

(4) (diversity premise) Some firms do pay for the predictions of economic theory.

Therefore

(5) (conclusion) (1) Ceteris paribus, firms that are paying for the predictions of economic theory are purchasing predictive information that is in general more accurate than any cheaper available alternatives.

(6) (conclusion) (2) Ceteris paribus, economic theory makes some predictions that are in general more accurate than any cheaper available alternatives.

The population here consists of firms again. The property some firms have is not paying for the predictions of economic theory. The further advantageous property that may follow is lower costs. The criteria in terms of which agents would do better is net revenue.

So we have an instance of an arbitrage argument. But note that the conclusion here is in effect the adaptation premise (4) in (RE), and it is drawn by denying that all firms avoid paying for the predictions of economic theory rather than by asserting the antecedent of the fitness premise (2). (PS) is problematic because its fitness premise (2) is dubious. Without knowing the causal pathways that make firms that hire economists viable, it is hard to know whether hiring economists might benefit firms in some other way instead.

To illustrate how this might be, consider a parallel argument that one might suppose some medieval proto-economist to have offered. To make this argument, just substitute in the last argument the word "kings" for the word "firms" and "astrological theory" for "economic theory". The evident conclusion is that astrological theory made better predictions than its medieval competitors. How else could kings that employed astrologers have survived?

The argument for astrology strikes almost everyone in much the way that Hands' argument strikes most non-economists. Those who would defend the predictive success argument for economics and not for astrology, must distinguish between the two arguments. One might, for example, argue that relations among different kingdoms were out of equilibrium, but surely so are relations among firms. One might argue that possessing an astrologer can be valuable to a king with a superstitious population, even if astrological theory is predictively worthless. But an economist can, of course, be valuable to a firm with a superstitious group of stock holders in just the same way. One might also feel uneasy about detaching the ceteris paribus clause from the conclusion. Kings who employed astrologers might also have been more wealthy or enterprising than those who did not. But firms that hire economists are not obviously the same in all other relevant respects as are those that do not. These qualms about the argument for the predictive worth of astrology should also apply to the argument for the predictive worth of economics. Without having to attend to the actual causal history, it is easy to let one's prejudices hide how unfounded one's premises are.[6]

My second example is Becker's and Friedman's famous argument concerning racial discrimination by firms:

(RD: Racial Discrimination Argument) A businessman or an entrepreneur who expresses preferences in his business activities that are not related to productive efficiency is at a disadvantage compared to other individuals who do not. Such an individual is in effect imposing higher costs on himself than are other individuals who do not have such preferences. Hence, in a free market they will tend to drive him out. (Friedman 1962, pp. 109–10)

With some additions Becker's and Friedman's argument, (RD) fits into the general schema as follows:

(RD′)

(1) (existential premise) Some firms will hire workers without regard to race.

(2) (fitness premise) If a firm that hires workers without regard to race has lower labor costs, then ceteris paribus it will have higher net revenue and will tend to displace those that

make smaller profits (or force them to change their ways).

(3) (equilibrium premise) Firms that hire workers without regard to race are not for the most part outperforming those that do not.

(4) (adaptation premise) Ceteris paribus firms that hire workers without regard to race have lower labor costs.

Therefore

(5) (conclusion) Ceteris paribus firms all hire workers with more or less no regard to race (i.e., there is no racial discrimination in hiring).

This argument fits the general schema nicely. The population consists of firms. The property some have is not discriminating against black workers. The further property this one leads to is lower labor costs. The criterion according to which firms with lower labor costs do better is net revenue or general competitive success. Given the existence of equilibrium, it follows (ceteris paribus for the most part) that there is no racial discrimination.[7]

I have taken some liberties here with what Friedman says, in order to display clearly the formal similarities with other arbitrage arguments. Friedman may not intend such a strong argument. One might only argue that *eventually* the workings of the competitive market will insure that there will be no discrimination in hiring. Unless the system moves rapidly to equilibrium, this weaker conclusion is consistent with systematic racial discrimination.

Let us focus on the stronger form of the argument, regardless of whether Friedman or Becker intended to make it. Is it sound? The existential premise (1) seems obvious: some firms do not discriminate on the basis of race. (2), the fitness premise, *seems* just as obvious: If not discriminating lowers labor costs, then, ceteris paribus, it increases net revenue, and firms with higher profits will, ceteris paribus, displace firms with lower profits or force them to change their ways. (4) is plausible as well. And, finally, to the best of my knowledge, the equilibrium claim (3) is true: non-discriminatory firms are not for the most part displacing those that discriminate or forcing them to change. Yet the conclusion is obviously false.

Where are the unobvious difficulties with the premises? Just where one would expect – in the adaptation and fitness premises and their ceteris paribus clauses. As Becker himself points out, there are circumstances in which one should either reject (4) or not that its ceteris paribus conditions is not satisfied: Hiring workers without regard to race will not lower one's labor costs if workers of different races do not work well together. Furthermore, the ceteris paribus clause in (2) may never be satisfied. As Akerlof persuasively argues, hiring black workers can considerably increase *non*-labor costs, given racial prejudice on the part of only a few suppliers or customers. Such costs would be obvious if one examined the actual causal history that purportedly forced firms to give up their discriminatory ways or that purportedly drove the recalcitrant out of business altogether. Without such study or some good substitute, the ceteris paribus clauses in the fitness and adaptation premises will be unjustified guesses.

The strength of sound arbitrage arguments is that they establish that a population or all of its members must have certain traits merely because of the existence of some members with certain traits and a competitive environment. Thus they avoid having to make assumptions about given universal traits of human beings and concerning particular causal paths.[8] If ceteris were paribus and hiring without regard to race really did lower total costs, then the market would lead toward the elimination of racial discrimination in hiring whether or not a great many employers were racist. Indeed, one apparently needs almost no assumptions at all about the motivation of firms. Those that do not act in order to maximize profits, or hire with regard to race, or make predictions about the future that do not match those of economic theory, and so forth, will simply be eliminated. But the world is a complicated and messy place, and if we do not attend to what is actually going on, our arguments will all too often turn out to be the venting of our prejudices instead of our insights into nature's secrets. If we want knowledge of matters of fact concerning various populations and their members, we're going to have to do empirical research. Even the best attested scientific theory provides few shortcuts here.

Arbitrage arguments only work when their premises are true and the ceteris paribus conditions can be met. Thus they will, in my view, have little application to consumer behavior, where the fitness premise will rarely be satisfied,[9] and even application to firms is dubious. Alchian's and Friedman's argument that firms must behave as if they were profit maximizers has already been forcefully (decisively in my opinion) criticized by Winter (1964). The argument for the rational expectations hypothesis is a relatively strong one, although the presumed predictive superiority of economic theory and the extent of the fitness provided by this superiority can certainly be questioned. The Friedman-Becker argument concerning racial discrimination is mainly of value because the falsity of its conclusion demonstrates the falsity of its premises or the failure of their ceteris paribus clauses.

In closing let me add a few meta-methodological remarks. I take it as a truism that good philosophy of science must pay careful attention to what workers in particular scientific disciplines are doing, but that philosophy of science nevertheless remains distinct from history of science. This paper puts some flesh on this truistic skeleton as follows: Although possible, it would be difficult for a philosopher to recognize the sort of theoretical strategy manifested by arbitrage arguments. Only familiarity with the use of such arguments enables one to appreciate their strengths and weaknesses. Yet the training and concerns of philosophers are essential too – the knowledge and interests that lead to the quasi-formalizations above have a central role; and the ability and willingness to look beyond a particular discipline is also helpful. Finally a normative concern with good argument and the search for knowledge are central to the philosophical project.

Let me end by quoting a particular robust arbitrage argument. I'll leave its criticism to others:

> (SS Self-Serving argument) Methodological scrutiny and philosophical reflection on economics must be valuable. Otherwise those who avoid it would be more prominent, earn higher salaries and teach more students. Over time those who persist in such scrutiny and reflection would either see the light, be driven from the profession, or would simply die out without replacements. Since such a process is not in evidence, the value of methodological scrutiny and philosophical reflection is established.

NOTES

* I would like to thank Michael McPherson and Jonathan Pressler and members of the audience at "Philosophy of Science II", especially Cristina Bicchieri, Bert Hamminga and Philippe Mongin, for helpful comments.

[1] Actually, as Cristina Bicchieri pointed out in discussion, one is forced to a stronger conclusion. For if agents are rational and, on average, their expectations match those of economic theory, then this information, too, will be used and the expectations of individuals will collapse to the average expectations. In more recent work, it is assumed that everyone's expectations are rational. Philippe Mongin pointed out that this argument contains itself in one of its premises – see premise 4 of the reformulation. This self-reference reduces the argument's persuasive power, although it does not, of course, render it invalid.

[2] Ignoring the ceteris paribus and "for the most part" qualifications, and simplifying, one can rewrite the argument in the first order predicate calculus as follows:

(1)	$(Ex)Hx$	existential premise
(2)	$(x)\{[(Hx \rightarrow Ax) \,\&\, \sim(x)Hx] \rightarrow Fx\}$	fitness premise
(3)	$(x) \sim Fx$	equilibrium premise
(4)	$(x)(Hx \rightarrow Ax)$	adaptation premise

thus

(5) $(x)Hx$

where the domain of quantification is restricted to firms, "*Hx*" should be read "*x* has expectations matching the predictions of economic theory, "*Ax*" should be read "*x* has more accurate expectations than firms whose expectations do not match the predictions of economic theory", and "*Fx*" should be read "*x* makes larger than average profits". In simplifying the formal structure I have made use of a stronger equilibrium premise (3) and have omitted reference to indicators.

[3] One must be careful concerning the notion of "accuracy". Predicting that the weather tomorrow will be the same as it is today may be more accurate than are professional weather forecasts in the sense of frequency of correct predictions, but such predictions will, of course, miss all weather changes; and the losses from such errors may be much larger. To speak sensibly about accuracy requires some specification of the loss function for error.

[4] Here is a formal restatement of the schema. Although it involves second order quantification, particular arbitrage arguments need not:

(1)	$(e)(H)(Ex)He, x$	existential premise
(2)	$(e)(H)(x)\{[(He, x \rightarrow A_e x) \,\&\, \sim(x)He, x] \rightarrow_{\text{c.p.}} Fe, x\}$	fitness premise
(3)	$(e)_g \sim Fe^*, x$	equilibrium premise
(4a)	$(x)(H^* e^*, x \rightarrow_{\text{c.p.}} A_{e^*} x)$	adaptation premise

or

(4b)	$(Ex) \sim H^* e^*, x$	diversity premise

thus

(5a) $(x)_{\text{c.p.},g}\ H^* e^*, x$

or

(5b) $\sim_{\text{c.p.},g}(x)(H^* e^*, x > A_{e^*} x)$

I left implicit the restriction of H to $\{H(e)\}$ and of x to some population P. The subscript "c.p." abbreviates "ceteris paribus" and the subscript "g" abbreviates "generally, or for the most part". Again I have oversimplified the equilibrium premise (or placed an additional burden on its ceteris paribus clause to make the logical structure as evident as possible.

[5] One might, for example, offer an arbitrage argument to the effect that a taboo on eating cows must be advantageous (Harris 1966) or to the effect that a particular low energy molecular state must be universal within a given gas sample.

[6] This comment on Hands' argument was prompted by Alex Rosenberg's reaction:

> Hands's argument that economic theory is predictively successful because business and government would not continue to demand, pay for, and consume it if it weren't, has all the merit of an equivalent argument for alchemy. Since alchemists were in demand for over half a millenium, they must have had something worth selling. Survivability arguments have a place in economics.... But unguarded ones like this simply give them an undeservedly bad name. (1986, p. 130)

[7] Note that these cost advantages are available even to employers who hire and compensate their workers without regard to race at all and who pay just as well as the firms that do discriminate. For such employers will be able to attract better workers for the given wages. Note that Alchian's and Friedman's early arguments (1950, 1953) to the effect that firms act as if they were profit maximizers can also be seen as such arbitrage arguments.

[8] Sober argues, in fact, that what he calls "equilibrium explanations", which would include arbitrage arguments are not causal at all (1983). Although not specific about the particular causal mechanism, they nevertheless seem to me to be causal explanations.

[9] Thus consider an argument such as the following:

> (CM) Consumer mastery argument: Bayer aspirin must be better than its competitors, since people are willing to pay more for it. If it were not better, those people who bought cheap aspirin would feel just as well and have more money to spend. Others would learn from their example, or, through some experimentation would learn the lesson themselves. Bayer could not continue charging more for its aspirin unless it really were superior.

The advantages enjoyed by those who do not purchase Bayer aspirin if it were the case that Bayer were no better than other aspirin are weak and uncertain, and only a flimsy mechanism is suggested whereby the competitive outcome is supposed to be a drop in the price of Bayer aspirin or its elimination from the market.

REFERENCES

Akerlof, G.: 1985, 'Discriminatory, Status-based Wages among Tradition-oriented, Stochastically Trading Coconut Producers', *Journal of Political Economy* **93**, 265–78.

Alchian, A.: 1950, 'Uncertainty, Evolution and Economic Theory', *Journal of Political Economy* **58**, 211–21.

Becker, G.: 1957, *The Economics of Descrimination*, University of Chicago Press, Chicago.
Elster, J.: 1979, *Ulysses and the Sirens: Studies in Rationality and Irrationality*, Cambridge University Press, Cambridge.
Fisher, R.: 1931, *The Genetical Theory of Natural Selection*, 2nd. rev. ed., Dover, New York, 1958.
Friedman, M.: 1953, 'The Methodology of Positive Economics', in: *Essays in Positive Economics*, University of Chicago Press, Chicago, pp. 3–43.
Friedman, M.: 1962, *Capitalism and Freedom*, University of Chicago Press, Chicago.
Hamilton, W.: 1967, 'Extraordinary Sex Ratios', *Science* **156**, 477–88.
Hands, D.: 1984, 'What Economics Is Not: An Economist's Response to Rosenberg', *Philosophy of Science* **51**, 495–503.
Harris, M.: 1966, 'The Cultural Ecology of India's Sacred Cattle', *Current Anthropology* **7**, 51–59.
McCloskey, D.: 1986, *The Rhetoric of Economics*, University of Wisconsin Press, Madison.
Muth, J.: 1961, 'Rational Expectations and the Theory of Price Movements', *Econometrica* **29**, 315–37.
Rosenberg, A.: 1986, 'What Rosenberg's Philosophy of Economics Is Not', *Philosophy of Science* **53**, 127–32.
Sober, E.: 'Equilibrium Explanation', *Philosophical Studies* **43**, 201–10.
Winter, S.: 1964, 'Economic Natural Selection and the Theory of the Firm', *Yale Economic Papers* **4**, 255–72.

Manuscript received 25 January 1988

Carnegie Mellon University
Pittsburgh, PA 15213
U.S.A.

ALAN NELSON

AVERAGE EXPLANATIONS*

ABSTRACT. Good scientific explanations sometimes appear to make use of averages. Using concrete examples from current economic theory, I argue that some confusions about how averages might work in explanations lead to both philosophical and economic problems about the interpretation of the theory. I formulate general conditions on potentially proper uses of averages to refine a notion of *average explanation*. I then try to show how this notion provides a means for resolving longstanding philosophical problems in economics and other quantitative social sciences.

Good scientific explanations sometimes appear to rely on facts about averages. Sometimes this is because of the unavailability of data for the individuals being averaged over, and sometimes it is because the explanation is only good "on average" and inapplicable at the more fundamental level of the things being averaged over. In this essay, I shall try to refine a concept of *average explanation* by examining both unsuccessful and potentially successful attempts to use facts about averages to provide scientific explanations. Explanations that appear to make direct use of averages are especially common in the social sciences. Even in economics, which is often believed to provide explanations of the social behavior of individuals, frequent reliance on averages is often taken to be unavoidable. I shall focus attention on how average explanations are used in economics and argue that some of these uses have important consequences for the interpretation of central parts of economic theory. The arguments given can easily be generalized to theories in other social sciences that employ quantitative explanations. The economic theory of consumer demand provides the best social scientific case study for the philosophical points I want to make. Some explicit references will also be made to economic models employing the Rational Expectations hypothesis, but I shall not lay out the details of extensions of the essay's theses to other sciences.

Demand theory, the study of how consumer demands for goods varies with prices, is central to modern neo-classical economics because its potential for empirical usefulness, its intrinsic theoretical importance, and its foundational role in general equilibrium theory.

Erkenntnis **30** (1989) 23–42.

Empirical studies focus on estimating equations describing markets. An extremely simple example is

$$(1) \qquad Q^j = a^j\bar{y}/p^j + \epsilon^j \qquad \forall j$$

where Q^j is the market demand for the jth good, a^j is a constant, p^j is the price of the jth good, $\bar{y}$ is the average income, and ϵ^j is an error term with an appropriate distribution. (1) is far too crude to be of much use in empirical applications, but it captures most of the essential features of more sophisticated models.[1] The market behavior described by (1) could arise from individual behaviors actually conforming to

$$(2) \qquad q_i^j = b_i^j y_i/p^j \qquad \forall i, \forall j$$

where q_i^j is the ith consumer's demand for the jth good, b_i^j is a constant, and y_i is his income, or

$$(3) \qquad u_i(x^1, \ldots, x^n) = \sum_{j=1}^{n} b_i^j \ln x^j \qquad \forall i$$

where u_i is the ith consumer's utility and x^j is consumption of the jth good. In this example, we might substitute the average income for y_i in (2) and write

$$(4) \qquad \bar{q}^j = c^j\bar{y}/p^j \qquad \forall j$$

and call (4) the "average individual demand function". Similarly, we might find a utility function

$$(5) \qquad \bar{u}(x^1, \ldots, x^n) = \sum_{j=1}^{n} a^j \ln x^j$$

that yielded (4) when maximized subject to the budget constraint

$$\sum_{j=1}^{n} p^j x^j = \bar{y}$$

and call it the "average utility function". This might be defended by saying that if every consumer had (5) as his utility function, then they would all have the demand function (4) and the market demand would be the individual demands given by (4) multiplied by the number of consumers in the market. Economists often say that a utility function like (5) "rationalizes" an average individual demand like (4) which, in turn, rationalizes the market function (1).[2] A plausible next step would be to say that the average demand or the average utility function plays a role in *explaining* the average demand function. This is an instance of the notion of average explanation I am concerned with. The procedure might also be called *representative explanation* or "explanation by proxy", because if it were the case that every consumer were accurately described by the properties of the average consumer given by (4) and (5), the market demand relationship (1) would again be realized. It is important to note that representatives in this sense are not arbitrarily chosen individuals. On the contrary, a representative must be carefully chosen (or better, *constructed*) so that if every individual were described by exactly the same relationships as the representative, the aggregate relationship would be (1).

Another way of characterizing this notion of average explanation comes from the more practical problem of obtaining a functional form to econometrically estimate for the market demand, instead of by the more theoretical problem of explaining or rationalizing a given aggregate relationship. We might say, "*Begin* by assuming that the *representative* individual has (5). Now maximization gives (4) and summation gives (1). If (1) yields good econometric results, we will consider ourselves lucky and say that (5) was a fortunate heuristic". Now we might, again, consider ourselves to have an explanation of (1)'s success in terms of the averages given by (4) or (5). Incidentally, representative individuals often also play a crucial role in the construction of so-called Rational Expectations models. A common procedure is to assume that a representative consumer or firm will use available information fully rationally – they do at least as well as they would if they knew the best economic theory. Theorists can then take the aggregate demand and supply functions to have the same form as the representatives.[3]

The kinds of average explanation being envisioned for demand theory are of philosophical interest because they engage nearly all of the most important foundational problems in economic science. I shall concentrate on the connection between analyses of economic aggregates and analyses of individual economic agents. This topic introduces a wide range of problems, but I am presently concerned with the nature of economic explanation. Two questions about explanation that arise in microeconomics, for example, are: "Does the theory explain the behavior of individuals in addition to the behavior of entire markets?" and "Must the explanations of aggregates be interpreted as being somehow based on individual behavior?". It is uncontroversial that microeconomics is intended to provide explanations of aggregative phenomena such as relative prices. There is not much active controversy among economists over whether the theory explains individual behavior, but there certainly is much division of opinion on the matter. I shall try to clarify both sides of this matter and then determine how details about average explanations can be brought to bear on it.

There are many arguments for the conclusion that the behavior of actual individuals (or, more weakly, of representative individuals) is part of the subject matter of microeconomics. One argument is based on prima facie evidence. When one surveys texts and journals, one initially gets the clear impression that microeconomics is not intended to be *exclusively* about phenomena or quantities that are essentially aggregative. For example, when Kenneth Arrow considered the matter directly he wrote,

> In most mathematical and, generally, in most deductive studies in the social sciences, the starting point is the behavior of the individual.... There is one equation for each individual.....
>
> This individualistic viewpoint, as we may term it, is explicit in the main tradition of economic thought... (1968, pp. 640–1)
>
> ...[one] methodological principle emerges clearly: in order to have a useful theory of relations among aggregates, it is necessary that they be defined in a manner derived from the theory of individual behavior. In other words, even the definition of such magnitudes as national income cannot be undertaken without a previous theoretical understanding of the underlying individual phenomena. (p. 642)

This expresses a natural and attractive idea. Since social phenomena arise in regular ways from the behavior of individuals, one obtains the most powerful kind of social explanations by rigorously deriving them from good explanations at the level of individuals. The idea is usually put even more explicitly; one could build a formidable heap of all the books and articles that *define* the microeconomic level as the analysis of individual decision makers. Two respectable sources chosen from literally hundreds are the first chapter of Quirk and Saposnik (1968, "The Setting: Individual Economic Agents") and Frank Hahn's declaration: "I am a reductionist in that I attempt to locate explanations in the actions of individual agents" (1984, p. 1).

A second argument comes from realizing that the economic

behavior of particular human beings is an intrinsically important matter of great interest. In addition to the obvious scientific interest of this class of phenomena, there are philosophically important normative questions about what individuals *should* choose to do given that they do, or want to, behave rationally. What is the best method for studying these issues? It *might* be biology, neurology, psychology, artificial intelligence, or even armchair philosophy, but the obvious first attempt is to use economic theorizing. If empirical studies convince us that other methods do a better job, then we shall assign completely different tasks to microeconomics. There are presently no such studies. Nor are there any convincing methodological arguments against attempting to use the theory to explain individual behavior.[4]

Another strong argument proceeds from the belief that economic behavior is *essentially* rational. There is an inclination to think that any adequate scientific treatment of economic phenomena must depict them as outcomes of primarily rational behavior. It is difficult to comprehend how aggregative behavior could be rational in itself, intrinsically rational, without recourse to something metaphysically objectioanble like group minds. Therefore, agreeing that economics must be concerned with rationality leads one to require that it deal with clear manifestations of rational behavior in individuals. Large scale phenomena can then be regarded as rational in a derivative sense insofar as they can be derived from small scale elements. This is probably the strongest motive for interest in Rational Expectations and, in general, for attempts to provide microfoundations for macroeconomics. These enterprises do not make much sense if microeconomics does not include a theory of actual rational individual behavior.

Other persuasive arguments can be constructed, but these three suffice to make a strong case. It is perplexing, then, to discover that the arguments for the contrary position, that the theory is about only large scale phenomena, are also strong. The contrary position can be defended with positivistic or behavioristic doctrines which have it that the relevant facts about individuals are unobservable, or that the data that can be directly obtained for individuals are somehow unscientific. A better defense can be developed by maintaining that microeconomics is a predictive success for some levels of aggregation, but a predictive failure for actual individuals. If this is correct,[5] it would be odd to maintain that microeconomics is about the domain where it is an empirical failure and not about the domain where it is an empirical success.

The rest of this essay is an investigation of the relationship between average explanations and the competing interpretations of the scope of microeconomics. I hope to elucidate both issues. I shall first evaluate the extent to which average explanations may serve to extend the scope of microeconomics to individuals.

The problem with thinking that the scope of the theory is wide enough to include individuals is another of the classical conundrums in economics and in the social sciences generally, namely, the link between theory and evidence. Social scientific theories do not predict, retrodict, or explain as well as the best theories in the natural sciences. This is often attributed to the difficulty of collecting good data, or of experimentally controlling phenomena in order to generate good data. In any case, the fact of the matter is that there is very little usable, scientifically collected information about the economic behavior of particular individuals. There are, on the other hand, plenty of good aggregated data. Can they be utilized to construct average explanations whose explanandums include individuals? For example, might (4), a demand function for an average individual serve to explain or help explain the behavior of some real individuals? That might suffice to extend the full explanatory power of the theory to individuals. If so, we need an account of the relationship between the average and the particular individuals that enables a theoretical treatment of the former to be an informative, useful explanation of the latter.[6] Let us begin by looking at some very simple cases that seem to exemplify acceptable explanations of individual properties in terms of information about averages.

EXAMPLE 1. Suppose that I am quite ignorant of Northern European weather patterns and recalling home in Los Angeles I ask: "Why isn't it thirty degrees Celsius today here in Amsterdam?", or "Why is it raining here in the summertime"? An appropriate scientific explanation of these phenomena for me would include reference to the average temperature (about 20 degrees) or to the average number of sunny days for Amsterdam in July (few). There are, of course deeper explanations of these phenomena. I could be shown meteorological maps for the last week. And there is some objective sense in which this is a better explanation than an appeal to the average temperature.

There are, however, contexts like this one in which the appeal to the average seems to be a fully adequate answer, and perhaps an even more appropriate answer, to the why question even when a scientific answer is expected.

EXAMPLE 2. Here we have an almost completely unscientific example. Suppose I am an avid baseball fan and have memorized the information in Figure 1. A friend unfamiliar with Team A asks me why player #5 hit 0.290 for the month of June – perhaps he is thinking that #5, an older player, ought to be hitting less well. As in Example 1, it seems that the answer, "Because the team average is 0.295 and #5's 0.290 is very close to that", provides some kind of explanation. This explanation could be elaborated by pointing out that most of this team's players are hitting around 0.295, the owners fires anyone who is hitting below 0.275 unless he is a remarkable fielder, this team leads the league in hitting because they take extra batting practice, and so on. Again, there are deeper and better explanations having to do with details about #5 and who is hitting before and after him in the lineup and so forth. A full explanation of why the individual averages are distributed about the team average in the way that they are would also be relevant and informative. But my friend has received a reasonable, explanatory, answer to his why question.

Why do the appeals to averages in Examples 1 and 2 have whatever explanatory power that they do? The summer sky in Amsterdam (I am assuming) is usually cloudy; fine days are comparatively rare. Almost

Example 2		Example 3		Example 4	
Player 1	0.200	Player 1	0.270	Player 1	0.190
Player 2	0.289	Player 2	0.275	Player 2	0.190
Player 3	0.290	Player 3	0.280	Player 3	0.190
Player 4	0.294	Player 4	0.285	Player 4	0.190
Player 5	0.290	Player 5	0.295	Player 5	0.295
Player 6	0.296	Player 6	0.290	Player 6	0.400
Player 7	0.300	Player 7	0.300	Player 7	0.400
Player 8	0.301	Player 8	0.305	Player 8	0.400
Player 9	0.395	Player 9	0.355	Player 9	0.400
Team A average 0.295		Team B average 0.295		Team C average 0.295	

Fig. 1. Individual and team batting averages for three baseball teams.

all of the players in Example 2 are very close to the team average. It seems plausible in both of these cases that there is some underlying explanation of why these things are so. It is not a big accident that most of the individuals in these samples are near the mean values for the populations. I suggest that the explanations offered in Examples 1 and 2 operate by drawing one's attention to this fact. In each case the explainer says to the why-questioner, "Rest assured that the phenomenon you are asking about is what one would observe more often than not in these circumstances; it is to be expected. There is, moreover, a detailed, convincing explanation of why *that* is so, though I won't trouble you with it".

I think that this suggestion becomes clearer when we examine cases like the following that are similar in many respects but obviously do *not* have any explanatory power whatsoever.

EXAMPLE 3. Now my friend asks me why player #5 on Team B. is hitting 0.295. This time, I cannot rely on my knowledge to reply: "Because the team average is 0.295". The problem is that it is accidental that #5 is near (in this example, exactly at) the team average; the individual averages are almost uniformly strewn about. This time, the team average does not provide any evidence about #5's average. The attempt to give an average explanation fails just as badly for #5 as it would for #9 who is nowhere near the team average. In Example 2, however, it seems that appeal to the team average is explanatory for #5 because it would be explanatory for *most* of the players. Baseball is such that we strongly suspect some underlying common cause for the clustering of averages around the mean.[7] Intuitively, this makes a good condition on the explanatory power of the group average: it must be a good explanation for most of the individuals. Failure to meet this condition can results in cases where an appeal to the team average is worse than non-explanatory as in the next example.

EXAMPLE 4. My friend asks why Player #5 on Team C is hitting 0.295. Now the reply, "Because the team average is 0.295" is misleading in a way that is worse than no explanation at all. Suppose that the owner of the team is strongly committed to having only special kinds of players on the team. He wants only players who are extraordinarily good hitters or who are extraordinarily good fielders. Hitting ability is, alas, almost always negatively correlated with fielding ability. Hiring Player #5 was a mistake; in fact the owner intends to let him go as soon as his contract is up (#5 is a good, but not great fielder). So #5's average is completely unrelated to the team average which, therefore, can provide no evidence about it.

On the basis of these examples, I think we can conclude that successful explanations of the properties of particular individuals in terms of average properties must at least satisfy the following informally stated conditions:[8]

> *Condition* (i). The distribution of the individuals with respect to the property in question must have a principal region or interval of accumulation. In other words, most of the data points should be clustered together near the mean of the distribution. What counts as "most" depends on the details of the case in question and on our purposes in asking for an explanation.

Satisfaction of Condition (i) is clearly illustrated by a set of daily temperature readings that are normally distributed. Example 2 is an even clearer case. A skewed distribution, however, need not violate the condition as long as the population mean is within the region of greatest accumulation. Similarly, a multi-modal distribution might do provided that all but one of the modes are relatively small and the population mean is in the interval containing the principal mode. This leads to the second condition.

> *Condition* (ii). The particular individuals in question must be near the region of greatest accumulation. In other words, the individuals in question must be near the mean. What counts as sufficiently "near" depends on the details of the case in question and on our purposes in asking for an explanation.

The best case is obviously where the individual property to be explained has a measured quantity nearly equal to the population mean for the property. Example 4 is an illustration of what goes wrong when Condition (ii) is not met.

Thus, it seems that there are restricted circumstances in which appeal to average behavior has some limited amount of explanatory power. This is true even though facts about averages rarely give us the best explanations of individual behavior and often may not even provide very satisfactory explanations. I want to stress, however, that the main argument of this essay does not depend on this point. My argument is conditional: *if* appeals to averages are ever explanatory for individuals, then Conditions (i) and (ii) must be met. Furthermore, one could contest the claim that there are explanations even in Examples 1 and 2, but if there are *ever* good explanations that appeal to averages, then these seem to be good candidates.

Now we can ask whether there are average explanations in economics that satisfy Conditions (i) and (ii) so that they provide explanations of individual economic behavior. Strong explanations in the social sciences are hard to come by, so it does seem we should not scorn procedures that have explanatory power even when we know that better procedures may be possible in principle. Consequently, the answer to this question is of interest.

Let us begin by considering whether economic results such as (4) and (5) provide appropriate average explanations of individual behavior. A disanalogy between these and the successful examples given above is that, in economic practice, (4) and (5) are *not* obtained by tabulating all the relevant information for individuals and then calculating a mean. Instead, economists begin with aggregative information that enables them to assign values to the constants in aggregative relationships like (1). The specific features of (4) and (5), that is the constants a^j, *are calculated after the fact without the benefit of an investigation of any individuals. A logically* well motivated project would be to determine the values of the parameters in (2) and (3) and then aggregate these relationships to obtain (1). This is not done, partly for practical reasons and partly, I suspect, because no one really believes that equations like (2) and (3) describe actual people. This means that these average relationships are formulated in ignorance of how behavior may differ from individual to individual.[9] Hence, when it comes to determining whether we have a good average explanation of some individual behavior, we do not yet know enough to *determine* whether Conditions (i) and (ii) are satisfied. Is it *likely* they are satisfied? Condition (i) may be for some special commodities such as suntan lotion in Los Angeles. If that is so, then there must be many cases in which (ii) is satisfied as well. But for most commodities, I think it likely that individual quantities demanded are not distributed so as to meet (i). What about avocadoes? Some people like them and consume great masses of them. Others detest avocadoes and do not consume any. Is there a *typical* number (as opposed to a mean value) of avocadoes consumed? This is an empirical question, but in an unscientific investigation I discovered that there are very few indifferent consumers of avocadoes; people either have them regularly or not at all, the latter class being larger. That is reminiscent of Example 4 in which players averages had a peculiar bimodal distribution and average explanation was ineffectual. We have no reason to suppose that other ordinary commodities are any different in this respect: beer, pencils, gasoline, bicycle tires, and recordings of Brahms's music are all greatly demanded by some and little demanded by others. This is even more true for more "basic" commodities like wheat or iron ore – commodities for which derived demand, demand for their final products, predominates.

Similar results should be expected for the hypothesis that representative individuals have rational expectations. It is probably true that many of the most important wheelers and dealers in large markets deal with information efficiently, but it seems highly unlikely that *most* participants in most markets do so. Therefore, even if large scale relationships are such that they could result from aggregates of rational-expectations behavior, these relationships need not constitute average explanations of actual decision making processes.[10]

Since we are considering it to be economically *possible* that there be good average explanations meeting Conditions (i) and (ii), an a priori argument against their existence is ruled out. But the rudimentary empirical considerations I have mentioned provide strong evidence that there are no genuine explanations of individual behavior forthcoming from economic results about averages like (4) and (5) or about the kinds of representatives that appear in Rational Expectations models. I draw two closely linked conclusions from this. Accounts of individual behavior in which aggregative relationships (or derivative statements about averages) are supposed to serve as the explanans are not good average explanations. Also, it follows that no good argument for interpreting microeconomics as extending to individuals can be constructed from facts about average explanations.

Let us first explore the consequences of the conclusion about the scope of microeconomics. I indicated that there are behavioristic arguments in favor of refusing to theorize about the kind of human reasoning that contributes to the causation of economic behavior.

Those who maintain that microeconomics is exclusively a science of large scale behavior can also rebut the three arguments I gave for the contrary position. Against the observation that economists talk and write with individualistic language, it can be replied that this is done for pedagogical and heuristic reasons alone.[11] Against the fact that individual economic behavior is of intrinsic scientific importance, one might insist that it is the task of another science to study it.[12]

The reply to the individualist's third argument, that the best macro-explanation is derived from micro-explanation, is more complex. The point seems, at first, to be conceded by scientific anti-individualists (that is, those who are not concerned to say anything explanatory about individuals) like Milton Friedman:

> Both branches of theory [microeconomics and macroeconomics] analyze things in the small to further their understanding of things in the large: for example, the demand for cash balances by the individual holder of money, for monetary theory; the demand for bread or coffee ütensils by the individual hosehold, for price theory. (1976, p. 7)
>
> Our purpose in investigating the demand curve of the individual is to learn more about the market demand curve.... the individual decides somehow or other what goods and services to purchase. These decisions can be regarded as (1) purely random or haphazard; (2) in strict conformity with some customary, purely habitual mode of behavior; or (3) as a deliberative act of choice. On the whole, economists reject 1 and 2 and accept 3, partly, one supposes, because even causal observation suggests more consistency and order in choices than would be expected from 1 and more variation that would be expected from 2; partly, because only 3 satisfies our desire for an "explanation". (p. 35)

And Richard Lipsey writes:

> In practice we seldom have information about individual demand curves although we often do have evidence about the general slope of market demand curves. This derivation of marked demand curves by summing individual curves is a theoretical operation. We do this because we wish to *understand* the relation between individual curves and market curves. (1963, p. 63, emphasis added)

These economists are not interested in *explaining*[13] what the consumer of firm does. They are content to *assume* that individuals behave in accordance with economic rationality and use this *assumption* to help explain what the economist is really interested in: "things in the large", for instance, "market curves" or relative price levels. The idea must be that a *prior* knowledge of properties of individuals can help us to scientifically understand or explain corresponding properties of aggregates; if we have *no* reason to trust the assumptions about "things in the small", how could they further our understanding?[14]

This introduces a famous problem is economic methodology – the relevant assumptions about individuals are indeed false. If one is trying to learn about, understand, explain, or even predict the market demand for beer, it won't help to examine the beer demand curve of a teetotaler.

Many approaches to this problem have been proposed, but the basis for one interesting potential solution was proposed at least as long ago as Marshall:

> But the economist has little concern with particular incidents in the lives of individuals.... He studies rather "the course of action that may be expected under certain conditions from the members of an industrial group:, in so far as the motives of that action are measurable by a money price; and in these broad results the variety and the fickleness of the individual is merged in the comparatively regular aggregate of the action of many.
>
> In large markets then – where rich and poor, old and young, men and women, persons of all varieties of tastes, temperaments and occupations are mingled together – the peculiarities in the wants of individuals will compensate one another in a comparatively regular gradation of total demand. (Marshall 1890 [1961], p. 174)

This thought is repeated, often embellished with a reference to the Law of Large Numbers, in virtually every economics textbook published. Marshall realizes that it would be a mistake to use an economic fact about an arbitrarily selected individual (his demand curve, for instance) in an attempt to explain a corresponding fact about an aggregate (the market demand curve, for instance). It might seem possible, however, that Marshall's observation combined with what Friedman and Lipsey say about understanding an explanation suggests a way in which an *average* of individual demands can be explanatory. So we are faced with an effort to avoid a commitment to deal directly with varied and fickle people while retaining some of the micro-to-macro (in this case, individual-to-market) explanatory power that so deeply impresses the individualist. It is also interesting from this essay's point of view that the effort involves an appeal to a different strategy of *average* explanation, this time in the micro-to-macro direction. I argued that the reverse, macro-to-micro strategy does not produce good average explanations in economics. Now let us find out whether the present proposal succeeds in illuminating either average explanation or interpretations of economics.

If we had an independently well confirmed theory of individual demand behavior, it would be of great interest to show that accurate aggregative relationships could be derived from it. We do not have that theory. We have the pure theory of individual consumer demand which gives us relationships like (2), and we have empirical results like (4). I explained above why contemporary demand theory (and the whole of microeconomics for that matter) cannot be interpreted as a theory of the behavior of any particular individuals.[15] The principal problem is that we do not have *prior* determinations of the relevant properties of particular individuals. Furthermore, we have seen that conditions are not met for good explanations of individual behavior in terms of averages. Aggregative relationships like (1) are developed empirically, but relationships for representative individuals like (4) and (5) are constructed from the aggregates after the fact.[16] So, successful econometric models of market demand (insofar as they are successful) do not have their success *explained* by models of individuals or averaged individuals. It should also be pointed out that in economic practice, the estimation of aggregative relationships is rarely constrained by the theory of individual maximization. Economists want aggregative relationships that are empirically useful; a useful model that is not in accord with theory will not be readily abandoned and an empirically useless model that is rigorously derived from theory is of secondary interest. It is, of course, *possible* that the theory should prove an effective heuristic device in the construction of useful models. It is even possible that successful models of market demand should count as evidence in favor of some particular functional form for individual demands. But these possibilities have not been realized in the history of empirical work on demand. Demand theory has consistently lagged behind empirical work.[17,18] The epistemological priority of aggregates over individuals wholly confutes the strategy of *hypothesizing* that individual behavior is in accord with the theory in order to predict, explain, or understand aggregate behavior. Nothing is gained by these hypotheses. If we come to know useful relationships for aggregates by doing empirical work on aggregate data, then these relationships stand on their own. If they are not explanatory in their own right, then in the absence of microfoundations, they are not explanatory at all.[19] We must conclude that when microeconomics is interpreted as excluding explanations of individual behavior, there cannot be any good micro-to-macro average explanations – these require an independently confirmed theory of the individuals.[20]

So far, my results have been negative. I have not produced any examples of genuine average explanation in economics, nor have I been able to resolve the problem about the scope of microeconomics. I shall conclude with some remarks which I hope will prove constructive.

Since, as I noted at the outset, scientific explanations sometimes appear to make use of averages, this should be reconciled with the failure, to this point, to discover any good average explanations in economics.[21] Reconciliation can be effected by recalling that the characterization of average explanation as a relation between aggregative relationships like (1) and average relationships like (4) and (5) was quite restrictive. Averages can be informative in other ways. For example, equation (1) says that if average income increases by one unit, quantity demanded will *on average* increase by a^j/p^j units. If (1) were well confirmed, we might be inclined to extrapolate it to the future and predict that *on average* quantity will vary with price the same way it has in the past. Similarly, there is no problem in principle with extrapolating baseball Team A's team average for July to August. The same can be done at the level of individuals. If we had the data to run a good regression, we could use (2) to obtain analogous results for individual demands, on average.[22] Depending on facts about baseball, we may be prepared to do the same with the batting average for player #5 of Team A. Unlike the example of demand theory, these legitimate uses of averages do not attempt to link the micro-level to the macro-level; they operate properly only at a single level.

It is more difficult to say something constructive about the problem of determining the true scope of microeconomics. Individuals are as ubiquitous in microeconomics as they are empirically barren. What point can this part of the theory have if it is not to have an empirically adequate treatment of individuals or to have a deep, reductive explanation of aggregates? I shall tentatively propose an approach to answering this question.

In another essay, I argued that the primary purpose of abstract general equilibrium theory is not to *explain* or even predict phenomena, but instead to provide an existence proof for the observed phenomena.[23] The premisses of this existence proof are the ordinary assumptions about the individual level that we are loath to give up. The idea is to demonstrate the logical compatibility of economically rational individual behavior with an approximation of what is actually observed – some measure of market clearing and price stability in some Western economies. It cannot be denied that the general equilibria whose existence can be proved[24] are very rough approximations of real economies, but the purpose of the proofs, though important, is quite minimal.

People who are not thoroughly acclimated to working with neoclassical economic theory may find it surprising that self-interested economically rational individual behavior should lead to efficiency and stability. Beginning with neoclassical assumptions about individuals seems to establish a presumption in favor of the development of a Hobbesian state of war rather than a modern economy. A sceptic will declare that *even if* the microeconomic theory of individuals is developed to the point of empirical adequacy, it is unlikely that aggregative analysis can be given a foundation in this theory. An existence proof can serve the limited, but important purpose of defeating this sceptic. The possibility of a sceptical outlook means that few neoclassical economists will be completely satisfied with an aggregative theory that is demonstrably inconsistent with economically rational individual behavior, and that many neoclassical theorists tend to be uncomfortable with a theory that is even *possibly* so inconsistent.

This account of the cognitive significance of general equilibrium theory can be extended to partial equilibrium analyses and other branches of economics. Let us again consider demand theory. I have argued that (2)–(5) are, at present, empirically useless and that they do not serve to strengthen any explanations that (1) may provide. Consider, however, a sceptic about (1) who doubts the possibility of its fully specified form being generally consistent with individual utility maximizing behavior. This strong sceptical position is defeated by showing how (1) would result if all consumers had (5) or if (5) were the "average utility function" in some appropriate sense. Empirically specifying (2) would only be required to defeat a weaker sceptical position that as a matter of fact, in some given economy, (1) was not consistent with (2). Similarly, someone who maintained that business cycles were completely impossible given the assumption of rational expectation formation would find himself defeated by the "existence proofs" provided by some Rational Expectations models.

I conclude that we do not yet understand how social scientific facts about markets or more highly aggregated structures can actually be explained by theories of individual behavior. Explanations that appeal to averages of facts about individuals, initially the most promising approach, do not work in this context. In the course of establishing this point, conditions that must be met for *any* average explanation to work in the social or the physical sciences have emerged.

NOTES

* Some of the ideas in this essay were presented at the Philosophy of Economics II Conference at Tilburg University. I thank the participants for helpful remarks.

[1] The following equations (1)–(5) clarify the exposition because of their simplicity. Although they serve to illustrate the points I want to make, it has been many years since economists have used functional forms this crude in demand theory; references to sources discussing more sophisticated examples are provided below. Examples of refinements include taking some account of income effects by incorporating information about income distributions and price elasticities (the latter also reflect commodity substitution effects resulting from relative price changes). I shall sometimes write '(1)', '(2)', etc. for convenience when I mean to refer to a more sophisticated relationship between the variables. Instances of this will be clear from the context in which they appear.

[2] I shall continue to speak of individual economic agents even though equations like (4) and (5) are sometimes taken to describe households. For the purposes of this paper, the distinction is immaterial. In like manner, the reader should feel free to substitute "preferences described by a utility function like *X*" for "utility function *X*".

[3] For some examples distilled from the journal literature see Sargent (1979, especially Chapters XIII and XVI) or Sheffrin (1983). The device of the representative individual has an even longer history in the theory of the firm. The arguments in this essay apply there as well, but the extension will not be discussed here.

[4] This topic is taken up in some detail in Nelson (1986b).

[5] This point is incompatible with the last. Determining that microeconomics makes bad predictions for individuals would require *observing* that their behavior is in conflict with the predictions of the theory.

[6] I shall *not* argue that an "average person" is itself some special kind of individual and may, therefore, help us to explain actual people. Nor shall I try to conclude that microeconomics is *about* individuals (in some grammatical sense of "aboutness") from the premise that it is about averages which are themselves about individuals. Alexander Rosenberg has convincingly argued that typical scientific appeals to averages are about aggregates even when the grammatical form of the sentences expressing them might seem to obscure this (1976, pp. 34–45).

[7] In these successful examples deeper, better explanations require an account of why the relevant properties of the individuals are distributed the way that they are.

[8] One might try to sharpen this very informal presentation in various ways, but additional precision is probably not justified in these cases by strong intuitions of explanatory strength. Moreover, the treatment in the text suffices to make the points at issue.

[9] A clear representation of much of the relevant economic literature is provided in

Deaton and Muelbauer (1980, especially Chapter 6). There is also an interesting discussion in Händler (1980, pp. 147–150).

[10] For a cautious appraisal of attempts to econometrically model expectations see Sheffrin (1983, pp. 17–26).

[11] This reply is not very convincing. It is hard to understand how so many great economists (and not just textbook writing hacks) could produce a muddled treatment of the foundations of their discipline. I think that the best explanation of the divergent expositions is that they are reflecting divergent views. The reason this is not more widely recognized is that it is regarded as surprising to find disagreements about theoretical material appearing in introductory chapters in science textbooks. When there is disagreement about elementary matters, there is some tendency to think that some parties to the dispute are making a scientific mistake. This disagreement, however, does not reflect badly on economic science; introductory chapters in different mechanics textbooks also present conflicting views on how, for example, Newton's Laws are to be interpreted.

[12] One often reads that the study of preference *formation* is the province of psychologists. That is true, but it does not follow from that alone that all individual behavior, or even all cognition, is the exclusive province of psychology. The nature of economic decision-making by individuals *give* preferences or production functions could still be of interest to economic theory. Those who believe on principle that microeconomics excludes explanations of individuals thereby exclude all facts about individuals.

[13] Friedman is careful to enclose 'explain' in scare quotes. This device is a signal that he does not want to clutter his exposition with any philosophical mess. For some audiences, this is undoubtedly effective writing, but it should not escape our notice that Friedman helps himself to 'understand' and 'learn about' without the purifying scare quotes. These terms cannot be regarded as synonymous with the supposedly pure *predict* and are, therefore, just as philosophically loaded as 'explain'. Lipsey has also used 'understand' in the quoted passage.

[14] It would be interesting to investigate whether this position is fully compatible with individualistic social theories like Classical Liberalism.

[15] This is a statement about the current state of empirical economic knowledge and about prevalent economic methodology. An account of what would be required to extend current economics to individual behavior can be found in (Nelson, 1986b).

[16] In an earlier paper (Nelson, 1984) I argued that the inaccessibility of individual behavior might generally be an insurmountable obstacle to explaining macroeconomic relationships by providing microfoundations. Händler (1982) develops a formalism for some econometric models of market demand in which individual utility is eliminable in Ramsey's sense. In light of this he writes, "Individual utility theory plays no role in those fields where scientists aim at concrete explanation and prediction of human behaviour as well as finding instruments for influencing human behaviour" (p. 60). I think that this conclusion holds for all currently available models of market demand and that its validity extends to virtually all currently available models of aggregative phenomena that mention utility or profit maximization. Hal Varian seems only slightly less cautious:

> We have seen that consumer theory can contribute to the specification and estimation of demand systems. However, its contribution must be taken with a grain

of salt. The problem comes again in the fact that preferences may vary significantly from agent to agent.... [i]t seems that, if preferences can be regarded as distributed closely around some average representative utility function, aggregated demand behavior can be represented by neoclassical demand curves plus a random error component. (1984, p. 186–7).

[17] This is convincingly established in Stigler (1965) and in Deaton and Muellbauer (1980, Chapter 3. See also the references given in these sources to specific empirical studies).

[18] A better justified enterprise is the use of empirical results for markets to develope constraints on the parameters of equations that might describe represenattive individuals. (It is quite incorrect to conclude that individual demands will have exactly the same functional form as the market demand). This enterprise is, unfortunately, of purely theoretical interest. For a philosophical discussion of this issue see Nelson (1984, pp. 591–593).

[19] Deciding whether aggregative relationships can ever be explanatory in the absence of microfoundations that permit their derivation is a very complicated question that I shall not attempt to answer in this essay. A negative decision would provide a very strong argument in favor of extending the explanatory scope of microeconomics to individuals; otherwise, it is hard to see how microeconomics could be explanatory at all. It should be noted here that it is aggregated supply and demand that directly determine relative price levels. If it is true, as is sometimes said, that relative price levels (and their comparative stability) are the explanandums of microeconomic theory, there is still no practical, empirical requirement to derive these explanations from the actions of individuals.

[20] Again, the argument can be extended in the obvious way to potential micro-to-macro average explanations in the theory of the firm and to models incorporating the Rational Expectations Hypothesis.

[21] It is sometimes said that parts of physics like ideal gas theory provide an example of how macro-data can be used to explain the behavior of individuals. It is argued in (Nelson, forthcoming) that there are decisive disanalogies between such theories in physics and the economic theories under consideration.

[22] This must be contrasted with attempting to explain *average* behavior for a *particular* individual. Even though we might get a good explanation of the batting average ("average batting behavior") of #5 in Example 4 by doing a regression on data for him, we cannot explain his batting average by appealing to aggregative data (i.e., the team average).

[23] Nelson (1986a, especially pp. 171–175) introduces the idea and compares and contrasts it with technical presentations of existence proofs of general equilibrium, and existence proofs in other sciences.

[24] The standard treatment is Arrow and Hahn (1971). They also include some brief, but sensible methodological remarks.

REFERENCES

Arrow, K. J.: 1968, 'Mathematical Models in the Social Sciences', in M. Brodbeck (ed.), *Readings in the Philosophy of the Social Sciences*, Macmillan, New York.

Arrow, K. and F. Hahn: 1971, *General Competitive Analysis*, Holden-Day, San Franscisco.
Deaton, A. and J. Muellbauer: 1980, *Economics and Consumer Behavior*, Cambridge University Press, Cambridge.
Friedman, M.: 1976, *Price Theory*, Aldine, Chicago.
Hahn, F.: 1984, *Equilibrium and Macroeconomics*, MIT Press, Cambridge.
Händler, E. W.: 1980, 'The Role of Utility and of Statistical Concepts in Empirical Economic Theories: The Empirical Claims of the Systems of Aggregate Market Supply and Demand Functions Approach', *Erkenntnis* **15**, 129–157.
Händler, E. W.: 1982, 'Ramsey-Elimination of Utility in Utility Maximizing Regression Approaches', in Stegmuller, W. et al., eds.
Lipsey, R.: 1963, *Introduction to Positive Economics*, Weidenfeld and Nicholson, London.
Marshall, A.: 1890 [1961], *Principles of Economics* (9th ed.), Macmillan, New York.
Nelson, A.: 1984, 'Some Issues Surrounding the Reduction of Macroeconomics to Microeconomics', *Philosophy of Science* **51**, 573–594.
Nelson, A.: 1986a, 'Explanation and Justification in Political Philosophy', *Ethics* **97**, 154–176.
Nelson, A.: 1986b, 'New Individualistic Foundations for Economics', *Nous* **20**, 469–490.
Nelson, A.: (forthcoming), 'Human Molecules'.
Quirk, J. and R. Saposnik: 1968, *Introduction to General Equilibrium Analysis and Welfare Economics*, McGraw-Hill, New York.
Rosenberg, A.: 1976, *Microeconomic Laws*, University of Pittsburgh Press, Pittsburgh.
Sargent, T. J.: 1979, *Macroeconomic Theory*, Academic Press, New York.
Sheffrin, S. M.: 1983, *Rational Expectations*, Cambridge University Press, Cambridge.
Stegmüller, W., W. Balzer and W. Spohn (eds.): 1982, *Philosophy of Economics*, Springer-Verlag, Berlin.
Stigler, G. J.: 1965, 'The Early History of Empirical Studies of Consumer Behavior', in *Essays in the History of Economics*, University of Chicago Press, Chicago.
Varian, H.: 1984, *Microeconomic Analysis* (2nd ed.), Norton, New York.

Manuscript submitted January 25 1988
Final version received March 21 1988

University of California, Irvine
Department of Philosophy
Irvine
CA 92717
U.S.A.

ALEXANDER ROSENBERG

ARE GENERIC PREDICTIONS ENOUGH?

Predictive success is a necessary accomplishment of any discipline that claims to provide knowledge of a sort relevant to policy. Neoclassical economic theory does not provide much of this sort of knowledge, and this raises a serious problem, one which can be expressed in several different ways: What sort of a discipline is economics, if not a policy relevant science? How are we to interpret economic theory if we wish both to preserve it and absolve it of defects, in the light of its predictive poverty? Can the theory be revised or improved in ways that both retain its character, and make for predictive improvement? What form should a policy relevant predictively successful science of economic behavior look like?

All these questions involve the presupposition that economic theory is predictively weak. Similarly, they assume that economic theory's predictive strength cannot be improved, that whatever weakness is found in its application is due to economic theory itself, as opposed to, say, auxiliary hypotheses required to bring it into contact with economic data. Many will find this assumption obvious and beyond argument. Indeed, it is enshrined in a history of jokes about the subject: "if you put all the economists in the world end to end, they still won't come to a conclusion", etc. If anything, the claim that economic theory leaves much to be desired as a predictive science, should be treated as the received view about the subject. And the burden of proof should fall on those who hold either that it is not weak, or if it is, it nevertheless is improving.

1. FRIEDMAN ON THE PREDICTIVE WEAKNESS OF ECONOMIC THEORY

This burden of proof has sometimes been admitted by the most fervent defenders of neoclassical economics. It would be difficult to understand the point of Milton Friedman's famous paper, 'The Methodology of Positive Economics', (1953) except against the background of such an admission. At first blush it might be supposed

Erkenntnis **30** (1989) 43–68.

that Friedman's view is quite adverse to doubts about the predictive merits of economic theory. While on the one hand endorsing without qualification the claim that the goal of science is "valid and meaningful (i.e., not truistic) predictions about phenomena not yet observed," (1953, p. 7), on the other hand Friedman aims to vindicate economic theory as attaining this goal. But there is an important qualification in his vindication: for economic theory is deemed to have been predictively successful with respect "to the class of phenomena which it is *intended* to 'explain'". (p. 8, emphasis added).

But the question arises, what is the *intended* domain of economic theory? Along with other economists, like Hicks (1938), Friedman holds the intended phenomena to be facts about markets, industries and economies as a whole. By contrast, economics is not, according to Friedman, about individual economic choice, and therefore the failure of its purported predictions about such phenomena do not count in assessing its predictive success. This is the whole point of his famous attack on the relevance for assessing economic theory of testing its "assumptions" about individual expectation, preference and optimization, and the boundary conditions within which they operate. For it is a defence of neoclassical economics against the charge that economics is defective because its predictive record with respect to individual economic choice is lamentable, and the boundary conditions it stipulates are never realized. The defence consists in admitting the facts but denying their relevance to any assessment of economic theory.

Friedman's argument raises three questions: first are its claims about methodology in general sound; second are its claims about the predictive success of economic theory with regard to the allegedly "intended" domain of economic aggregates borne out; third, is the denial of any concern on the part of economic theory with individual behavior sustainable, or is it special pleading.

The first question I have dealt with at length elsewhere (Rosenberg 1976, pp. 155–170), as have many others. I shall say nothing more about it here. Let us turn to the second question. Friedman's evident aim is to defend economic theory against the claim that its assumptions are unrealistic. But why should this issue ever have arisen? Surely, the "unrealism", the idealized character of assumptions throughout the most successful theories of physical science has long ago settled the question of whether unrealistic assumptions are permissible in predictively successful scientific theorizing. The question of whether such assumptions are warranted in economic theory was not prompted by any general unease in the philosophy of science. It resulted from doubts about the predictive success of the theory for aggregate phenomena, and the desire to locate the obstacles to such success. This much appears to be clear in the very development in economic theory that seems to have motivated "The Methodology of Positive Economics".

Friedman writes: [p. 15]

> The theory of monopolistic and imperfect competition is one example of the neglect in economic theory of [Friedman's strictures against testing assumptions]. The development of this analysis was explicitly motivated, and largely explained by the belief that the assumptions of "perfect competition" or "perfect monopoly" said to underlie neoclassical economic theory are a false image of reality. And this belief was itself based almost entirely on the directly perceived descriptive inaccuracy of the assumptions rather than on any recognized contradictions of predictions derived from neoclassical economic theory. The lengthy discussion on marginal analysis in the *American Economic Review* some years ago is an even clearer... example. The articles on both sides of the controversy largely neglected what seemed to me the main issue – the conformity to experience of the implications of the marginal analysis – and concentrate on the largely irrelevant question whether businessmen do or do not in fact reach their decisions by consulting schedules, or curves, or multivariable functions showing marginal cost and marginal revenue. [p. 15]

But, as Blaug notes, (1978), "the original appeal of Chamberlain's book (*The Theory of Monopolistic Competition*) was that its predicted consequences were directly contrary to the implications of perfectly competitive models". Blaug gives an example: profit maximizers in perfect competition have no incentive to advertize. "However, advertising expenditures in an increasing number of product markets are a well-attested phenomenon" (1978, p. 416). This is British understatement at its best! But the theory of monopolistic competition predicts that producers of a heterogeneous product will advertise. Evidently historians of economics do not share Friedman's claims about what motivated studies of imperfect competition.

It might be argued, on Friedman's behalf, that the theory of imperfect competition arose from dissatisfaction about the Marshallian notion of the "representative firm" – surely an unrealistic idealization in the theory of perfect competition. But dissatisfaction with this notion stems not from its unrealism, but from the fact that such firms do not enjoy increasing or decreasing returns to scale. And the existence of this phenomenon is a fact economists wanted to predict and explain. The recourse to theories of imperfect competition reflected this aim, and not any intrinsic hostility to unrealistic assumptions. After all, the theories offered to account for imperfect competition had enough unrealistic assumptions of their own!

Much the same must be said about Friedman's other example. It is true that the marginal theory of the firm in a perfectly competitive industry, when combined with initial conditions about changes in demand, for example, makes some definite predictions about changes in supply and pricc, which other theories, with more realistic assumptions about the behavior of managers, do not make. But among these predictions at least half are disconfirmed, and the ones that are confirmed have a serious difficulty: economists have not seemed able to improve on them in any systematic way over the entire course of the history of the neoclassical theory of the firm. The confirmed predictions about markets and industries of this theory have at best always been "generic", or in Samuelson's terms "qualitative". When correct, they have told us the direction of a change, but there has been no tendency to go further, from the direction of a change to a well-confirmed statement of its amount or size or time-path. We shall return to the generic nature of economic predictions below. But the point here is that dissatisfaction with "the assumptions" of perfect competition or marginalism has been a *consequence* of dissatisfaction with the predictions of these theories, and not a direct source of discomfort to economists by themselves. To this extent Friedman's argument against rejecting unrealistic assumptions both attacks a straw man, and begs the question: no economist has ever questioned the assumptions of neoclassical theory *just* because they are unrealistic, and the real reason many have done so is because of dissatisfaction with what Friedman assumes to be beyond question, the predictive success of neoclassical economic theory.

This brings us to the third question: Friedman restricts the domain of predictions that permissibly test economic theory to *the* one "it is intended to explain" (p. 8). One wants to ask "intended by whom?" If Friedman, or any particular economist, is the authority on what the "intended" domain of economic theory is to be, then his thesis is altogether too easy to defend. Indeed, it becomes vacuous, for it is open to Friedman to simply reject any disconfirming test of any consequence of the theory as beyond *his* intended domain. More seriously, a cursory examination of the history of neoclassical theory shows that the intended domain of economic explanation certainly included the very phenomena described in the assumptions of neoclassical theory. Wicksteed, for example, opens *The Common Sense of Political Economy* (1910) with an explicit account of marginal utility as a literal explanation of individual human action, both economic and otherwise. He concludes the analysis thus: "Our analysis has shown us that we administer our pecuniary resources on the same principles as those on which we conduct our lives generally . . . in the course of our investigations we have discovered no special laws of economic life, but have gained a clearer idea of what that life is" [p. 126]. The fundamental assumptions of economic theory are psychological generalizations that explain individual choice and, by aggregation, economic phenomena.

In the course of the history of economic theory since Wicksteed and the marginalists, there has in fact been less and less emphasis on the explanation of individual choice as an aim of economic theory. But it remains to be seen whether this shift in the "intended" domain of explanation is anything more than an ad hoc restriction reflecting the disconfirmation of the theory with respect to such behavior. The shift from cardinal to ordinal utility and eventually to revealed preference certainly pruned economic theory's picture of individual choice. But this shift certainly appears to be one responding to the falsification and/or trivialization of successive versions of the theory of consumer choice. Revealed preference theory allows the economist to be almost entirely neutral on the psychological mechanism that leads to individual choice. But even this minimal approach makes implicit psychological assumptions, to wit, that the individual's tastes have not changed over the period in which we offer him pairwise choices over alternatives.

Economists may confidently announce, along with Hicks, that "economics is not in the end very much interested in the behavior of single individuals" (1939, p. 34), but this will not prevent false assumptions about individuals from bedeviling predictions about the economic aggregates made up of them. Similarly, thermodynamics may not be interested in individual molecules, but it is facts about these molecules that explain why the ideal gas law fails beyond a narrow range of temperature, pressure and volume. The neglect of economic assumptions can no more be justified by restrictions on the "intended domain" of economic theory than thermodynamics can be insulated from atomic theory. Or at least it can't if economics makes claims to being a theory with predictive consequences.

The upshot is that so far from constituting a defence of economic's predictive successes, Friedman's "Methodology of Economics" is rather a symptom of its weaknesses, both with regard to aggregate economic phenomena and individual economic behavior.

2. LEONTIEF ON THE SAME SUBJECT

Friedman's argument for the predictive success of economics turns out to be much less vigorous than it appears. But perhaps the demand that economic theory reflect substantial predictive success is too strong. After all, a discipline must walk before it can run: perhaps the most we should expect of neoclassical economic theory is *improvement* in predictive success. Suppose we can show that though the theory has not met with many predictive successes, it has met with some, and that the number, precision and significance of its successes is growing over time. If so, economists would, I think, have a right to be satisfied that their theory was on the right track, and they would with some justice be able to ignore complaints as wanting in patience.

There is of course a long tradition of criticism of economic theory that denies even this minimal achievement, indeed that alleges economists are not even interested in, as it were, walking, so that some day the discipline might run. Such criticisms are often founded on a more or less intuitive survey of the literature of economic theory. But sometimes they reflect allegedly hard data about the self-imposed insulation of economic theorists from the application of their theories to data. Perhaps the most persistent critic in this latter vein is Wassily Leontief, a Nobel laureate in economic theory, whose strictures can not be fobbed off as uninformed or splenetic.

Leontief does not argue that economic theory has made little progress in predictive power. He simply assumes it. In 'Theoretical Assumptions and Nonobservable Facts', his 1970 presidential address to the American Economic Association, Leontief attempts rather to explain why economic theory has shown only minimal improvement, and he condemns the discipline for its indifference to this fact. Leontief's assessment of the state of economic theory does not begin with any Luddite rejection of the highly mathematized character of the discipline. He does complain about the indifference to testing the assumptions of most mathematical models. He holds that "it is precisely the empirical validity of these assumptions (of formal models) on which the usefulness of the entire exercise depends. What is needed, in most cases, is a(n)... assessment and verification of these assumptions in terms of observed facts. Here mathematics cannot help, and because of this, the interest and enthusiasm of the model builder suddenly begins to flag: 'If you don't like my set of assumptions, give me another and I will gladly make you another model; have your pick" (1985, p. 274).

It is no surprise that this claim should fall on deaf ears among economists. For most have adopted Friedman's attitude with respect to assumptions. Moreover, even if Friedman's attitude is too complacent, Leontief's is too harsh. Unrealistic assumptions are not problematic in general, and the mere fact that economic theory begins with idealizations cannot by itself be the source of its difficulties.

But Leontief has a better argument for the same conclusion that economic theory shows no signs of improvement. Turning from mathematical models in the theory, he asks, "But shouldn't this harsh judgment be suspended in the face of the impressive volume of econometric work?" The answer, he writes, "is decidedly no". Advances in econometrics are in his view attempts to design statistical tools to extract significant findings from a data base that is weak and apparently not growing. What is more, "like the economic models they are supposed to implement, the validity of these statistical tools depends itself on the acceptance of certain convenient assumptions pertaining to stochastic properties of the phenomena which the particular models are intended to explain – assumptions that can be seldom verified. In no other field of empirical enquiry has so massive and sophisticated a statistical machinery been used with such indifferent results. Nevertheless, theorists continue to turn out model after model and mathematical statisticians to devise complicated procedures one after another" (1985, pp. 274–5).

So, it's not just the unreality of the assumptions of economic theory that results in its lack of predictive progress, it's also the fact that economists do not or cannot augment the data that will enable them to identify progress in prediction, nor have their econometric techniques proved powerful enough to derive results about the slim body of data already in hand. And, worst of all, there is no institutional motivation in the discipline to improve on this data base. For advances in formal modelling or the mathematics of statistical testing are more highly prized in economics than the augmentation of data.

One possible reply to this charge needs to be set aside immediately, though Leontief does not specifically address it. The reply is that the collection of data must be guided by theory. Without a fairly specific body of hypotheses to identify what variables are to be measured, observation will at best haphazard, and at worst pointless. Economic theory has perhaps not developed far enough to inform us as to what sort of data will test it adequately, and about what kinds of aggregate or individual phenomena it can be expected to predict. To this appeal for patience with economic theory Leontief might reply in terms of a claim he made twenty years before he wrote 'Theoretical Assumptions and Nonobserved Facts':

> If the great 19th-century physicist James Clerk Maxwell were to attend a current meeting of the American Physical Society, he might have serious difficulty in keeping track of what was going on. In the field of economics, on the other hand, his contemporary John Stuart Mill would easily take up the thread of the most advanced arguments among his 20-th century successors. Physics, applying the method of inductive reasoning from quantitatively observed events, has moved on to entirely new premises. The science of economics, in contrast, remains largely a deductive system resting upon a static set of premises, most of which were familiar to Mill and some of which date back to Adam Smith's *Wealth of Nations*. (1951, p. 15)

In short, economists have had the same theory in hand for upwards of two hundred years. The theory has not been increasingly accomodated to data it has led us to uncover, it has not even motivated much augmentation of data. Leontief's arguments can hardly be treated as expressions of impetuousness.

Leontief makes an important exception to his assessment of contemporary economics: agricultural economics. Here he says there has been a "healthy balance" between theory and application, and a resulting secular improvement in predictive accuracy. Without gainsaying these improvements, however, two things should be noted: first, many of the problems of agriculture to which economic thinking has been applied are "technological" – they are problems of constrained maximization of production or minimization of cost where the constraints are not economic, but physical, and are understood in great quantitative detail through the application of the "hard" sciences. The very factors Leontief sites, crop rotation, fertilizer effects, alternative harvesting techniques, are simple enough variables to bring the economy of the farm far closer to the assumptions of neoclassical theory than the economy of other units of production can be brought. This suggests that so far from holding out promise for the rest of economic theory, the success of agricultural economics helps us see its predictive limits. For beyond the solution of linear programming problems, economic theory itself has added little to what the agricultural sciences themselves can tell us about farm production. Secondly, agricultural economists themselves complain in terms very like Leontief's own, about the difficulty of increasing neoclassical economics' quantitative understanding of the operation of agricultural markets. Distinguished agricultural economists lament the over-development of formal theory and econometric techniques, and complain that the discipline does not adequately serve the needs of agriculture.

What agricultural economics has going for it is far greater completeness in the collection of data deemed to be relevant to agricultural policy. As much as anything the existence of these systematic and reliable data are responsible for the successes of the subject. But the data exist in large part because they are easier to secure, and far simpler to organize into homogeneous categories than the data that could test economic theory elsewhere. Moreover, the hypotheses about the data which are most strongly confirmed are, as noted above, technological, and not economic. One of the problems of non-agricultural economics is the difficulty of securing data. This has encouraged economists to concentrate on theory, instead of applications, and so engendered Leontief's problem. To begin to correct the situation, if it can be corrected, we need the right explanation of why the provision of these data is so difficult.

Twelve years after offering this assessment of the problems of economics, Leontief's views had not changed. In 'Academic Economics' (1982) he summarized a content analysis of the leading journal in economics. This analysis showed that over a ten year period the proportion of papers in the *American Economic Review* which elaborated mathematical models without bringing the models into contact with data exceeded 50%, and that a further 22% involved indirect statistical inference from data previously published or available elsewhere. The next largest category, approximately 15%, was described as "analysis without mathematical formulation and data".

Leontief concluded:

Year after year economic theorists continue to produce scores of mathematical models and to explore in great detail their formal properties; and the econometricians fit algebraic functions of all possible shapes to essentially the same sets of data without being able to advance, in any perceptible way, a systematic understanding of the structure and operations of a real economy. (1985, p. xii)

3. ARE GENERIC PREDICTIONS ENOUGH?

It's not that economic theory has no predictive power; rather the problem is that it does not have enough. And it never seems to acquire any more than it had at the hands of, say, Marshall in the late nineteenth century. We can illustrate this limitation by a glance at Paul Samuelson's *Foundations of Economic Analysis*. So far as imposing demands of testability and prediction on economic theory, this work is a latterday locus classicus. Samuelson's explicit aim was to formulate *operationally meaningful* theorems, by which he meant "hypotheses about empirical data which could conceivably be refuted. . . ." (Samuelson, 1947, p. 4). He continued:

In this study I attempt to show that there do exist meaningful theorems in diverse fields of economic affairs. They are not deduced from thin air or from a priori propositions of universal truth and vacuous applicability. They proceed almost wholly from two types of very general hypotheses. The first is that the conditions of equilibrium are equivalent to the maximization (minimization) of some magnitude. (The second is) the hypothesis . . . the system is in 'stable' equilibrium or motion. (p. 5)

What is worth noting here is the sort of empirical tests to which Samuelson insists economic hypotheses be exposed are not quantitative, but, as he calls them "qualitative":

In cases where the equilibrium values of our variables can be regarded as the solutions of an extremum (maximum or minimum problem), it is often possible regardless of the number of variables involved to determine unambiguously the *qualitative* behavior of our solution values in respect to changes of parameters. [p. 21]

But what exactly are qualitative predictions? Samuelson writes:

In the absence of complete quantitative information concerning our equilibrium conditions, it is hoped to be able to formulate qualitative restrictions on slopes, curvatures, etc., of our equilibrium equations so as to be able to derive definite qualitative restrictions upon responses of our system to changes in certain parameters. It is the primary purpose of this work to indicate how this is possible in a wide range of economic problems. (p. 20).

The qualitative predictions which we can test against data are roughly the signs, positive or negative, of the partial differentials of the changes in the values of economic variables we set out to measure. Thus, to take one of Samuelson's simpler examples, we may predict that an increase in the tax rate will lead a firm to a lower or higher output depending on whether the second partial derivative of output with respect to the tax-rate is positive or negative. And though we cannot reliably detect the actual quantitative value of this variable, we can more reliably determine its sign, that is determine whether the first partial derivative of output with respect to taxation is increasing or decreasing.

In general, qualitative predictions purport to identify the direction in which changes move, without however, identifying the magnitude of these directions. Of course, as Samuelson notes, we would like to have more than qualitative predictions

> ... purely qualitative considerations cannot take us very far as soon as simple cases are left behind. Of course, if we are willing to make more rigid assumptions, either of a qualitative or a quantitative kind, we may be able to improve matters somewhat. Ordinarily the economist is not in possession of exact quantitative knowledge of the partial derivatives of his equilibrium conditions. None the less, if he is a good applied economist, he may have definite notions concerning the relative importance of different effects; the better his judgment in these matters, the better an economist he will be. These notions, which are anything but a priori in their original derivation, may suggest to him the advisability of neglecting completely certain effects as being of a second order of magnitude ... In the hands of a master practitioner, the method will yield useful results; if not handled with caution and delicacy, it can easily yield nonsensical conclusions.

But unless we have an independent criterion of caution and delicacy, the difference between useful results and nonsensical conclusions will just be the difference between confirmed ones and disconfirmed ones. In the forty years after Samuelson published the *Foundations*, no such independent criterion has been forthcoming. But without it, or without well-substantiated numerical values for the parameters and variables of economic theory, neoclassical economics is condemned to *generic* prediction at best.

By generic predictions I mean predictions of the existence of a phenomenon, process, or entity, as opposed to specific predictions about its detailed character. Generic prediction is characteristic of most theories that proceed by establishing the existence of an equilibrium position for the system whose behavior they describe, and then claim it moves towards or remains at this equilibrium value. Classical examples of such theories are to be found in thermodynamics and evolution. (In fact Samuelson identifies his method with that of thermodynamics (p. 21).

Evolutionary theory, for example, cannot, in the nature of the case, predict the course of evolution, at most it informs us that over the long run, biological lineages will manifest increasing local adaptation to their environments. The theory tells us they are moving towards an equilibrium level of population, and of other individual traits that maximize their prospects for survival. For further details about the particular character of these adaptations, why they obtained instead of other equally advantageous alternatives, how they work to secure adaptation, etc., for all this, we need to appeal to other, non-generic, specific theories. Similarly, for the second law of thermodynamics, which informs us that energetic systems in the long run move towards an entropy-maximizing equilibrium, without telling us what course they take towards this equilibrium level. This lack of specificity is a weakness in generic theories. But it is an unavoidable concommitant of their explanatory strategy. A theory that explains behavior by claiming only that it always remains in or near some equilibrium is bound to be generic, for it must be consistent with whatever is done by the system whose behavior it describes.

But in general generic predictions are not enough. They are not a natural stopping place in scientific inquiry. And in other disciplines, generic theories have either been supplemented or improved, in order to acquire specific predictions. Thus, if we add theories about heredity, physiology, development, behavior and environment, to evolution's mechanism of variation and natural selection, we can hope to improve generic predictions into increasingly specific ones. And in thermodynamics, if we provide ourselves a measure of entropy, and a description of the mechanical and thermal properties of a system we can make specific predictions about the amount of entropy-increase it will manifest.

This is what we need to do in economics: either supplement the theory with theories from other disciplines that will enable us to convert generic into specific predictions, or find measures of the independent or exogenous variables of the theory that will enable us to do so. We need to do these things if economics is to justify any confidence as a policy relevant science. I suspect that this is the real moral of Leontief's lament for economics. Generic prediction is something, it is a start, but it is not enough. And economists should not be satisfied with it.

4. GENERIC PREDICTIONS IN KEYNESIAN MACROECONOMICS AND RATIONAL EXPECTATIONS THEORY

The shift from classical economic theory to Keynesian macroeconomics, and the rational expectations counter-revolution illustrate the degree to which economics as a discipline seems unable or unwilling to transcend generic predictions. The illustration's force does not rest on adopting any of these three theories, least of all the most recent, and it is only one of many examples. Its advantages for our purposes are that it reveals, in the context of most lively debate in recent economic theory, the role of prediction in economics, and the limitations upon it.

Classical economic theory predicted the existence, stability and uniqueness of a market clearing general equilibrium. True, the existence was not mathematically "nailed down" until the thirties, nor were stability and uniqueness established till even later. But, counting equations and unknowns seemed enough to substantiate the generic prediction of equilibrium. That the prediction was generic should be obvious. For business-cycle "fluctuations" could only be accommodated to the theory insofar as its assertions were not specific but generic. Indeed, business cycle theory was itself restricted to the generic explanation of the possibility of such fluctuations, and never reached the point of attempting to project their dates, durations, and magnitudes.

The depression and deflation of the nineteen thirties seriously undermined confidence in even the weak generic predictions of classical theory. To be sure, there were ways of reconciling the existence of a market-clearing equilibrium somewhere in the long run. But, as Keynes noted, in the long run we are all dead. So, Keynes developed a theory in which a non-market clearing equilibrium was possible, in particular one which did not clear the labor market, and so explained (generically) the existence of involuntary unemployment. This at any rate is the interpretation of *The General Theory* taught in most contemporary macroeconomic texts.

Keynes' theory is expressed in terms of a set of standard equations, laying out real national income as a function of consumption and investment (with these in turn functions of the interest rate and income), the demand for money balances as a function of income and the interest rate, and aggregate production, as well as the demand and supply of labor dependent on the wage-rate and the price level. The consequences of the theory for fiscal and monetary policy are given in the well-known "Hicks-Hansen" IS-LM curve. The IS curve gives all the values of national income and the interest rate which equilibriate planned investment and planned savings, whence the name IS. The LM curve reflects the set of points of income and interest rate at which the supply and demand for money balances are equal. The point at which these curves cross has the property of being a level of national income and interest rate at which planned savings and investment are equal, and there is no excess demand or supply of money. On the standard view of the shape of these curves this point is a stable equilibrium.

Two important consequences are usually drawn from Keynesian theory: first an explanation of how unemployment equilibria are possible, and second, a recipe for "fine-tuning" an economy to keep it close to a full employment equilibrium level of income. The first is often expounded in terms of how the shapes of the IS and LM curves can be affected by wage-rigidities, the phenomenon of the liquidity trap, and an investment demand curve that it is inelastic for changes in the interest rate.

The second implication, that by fiscal and monetary policy the government can shift the IS and LM curves to the right, in order to increase the equilibrium levels of income, and thus lower the level of unemployment, proved in some ways a more exciting consequence of the theory. For one thing, Keynes' explanation of the depression as a phenomenon of unemployment equilibrium was the explanation of a possibility. That is, it showed how something conventional economic theory deemed to be impossible might after all be possible. If, as many held, persistent high unemployment was not merely possible, but actual, then economic theory had to be reconciled with it. In establishing how unemployment equilibria were possible, Keynes preserved economic theory from the charge of irrelevance in principle. But neither Keynes nor his successors did much more than show this possibility. And the test of his theory's explanatory power was taken to be its implications for fine tuning the economies of industrial nations in the period after the depression ended. Here there were hopes of doing more than merely showing that what most people deemed to be actual was after all possible. The prospects of shifting the IS and the LM curve in order to "fine-tune" the economy was just what a discipline hankering for predictive success needed.

Suppose that the equilibrium values of national income, Y (on Figure 1) is below a full-employment level, and suppose that there is no budget deficit. Then by shifting the IS or the LM curves, or both, to the right, the government can raise the equilibrium level of income, and thus reduce unemployment. The IS curve is shifted to the right by increasing government outlays. This results in a new equilibrium level of income, Y', higher than Y, at the point of intersection of the LM and new IS curves. The government's new outlays of course produce a deficit, which can be made up either by selling bonds or increasing the money supply. Suppose the government sells bonds. This will raise interest rates above i, as the rightward shift in the IS curve indicates. But, on the Keynesian account of the matter, the rise in interest rates, which induces people to hold bonds, results in an increase in their

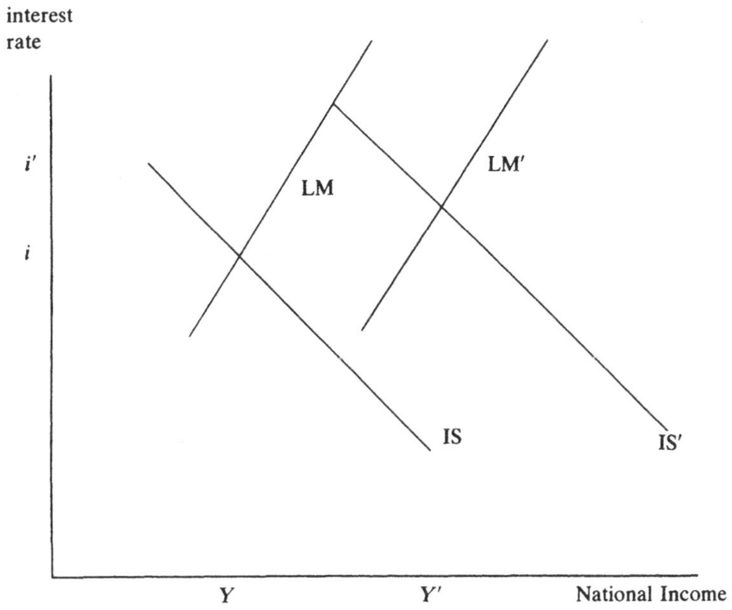

Fig. 1.

wealth, the bonds being counted as among their assets. Since people are now richer, they may spend more on commodities and services, therefore raising the level of national income, and shifting the IS curve even further to the right.

On the other hand, the government can finance the deficit by increasing the money supply instead of selling bonds. Increasing the money supply however shifts the LM curve to the right, thus moving the equilibrium level of national income even higher. The increase in real national income however, increases demand, and therefore reduces the real value of the money supply, i.e. produces inflation in the price-level. On the other hand, the increased national income produces higher tax-revenues and reduces the budget deficit. If all this happens at the right rates and in the right amounts, the end result is a balanced budget at a new and higher level of national income, and a lower level of unemployment. Note that the units on any IS-LM curve are purely notional, and the Keynesian predictions are purely generic. It is important to bear in mind that econometric models that attached "real numbers" to macroeconomic variables never unequivocally confirmed this ideal model.

Given the theory, fine-tuning was supposed to work. But it didn't. In particular, the economic phenomena of the seventies showed that the so-called Phillip's curve, tracing the trade-off between unemployment and inflation which results from increasing the money supply – i.e., shifting the LM curve to the right, either broke down or didn't hold at all. The generic predictions of Keynesian theory never had a chance to give rise to quantitative ones before they began breaking down. Ironically, having first shown how the actual was possible in the case of unemployment equilibrium, economics now had to show how the actual was possible in the case of "stagflation" – inflation with increases in unemployment.

The failure of fine-tuning, which seems to disconfirm the conventional Keynesian model's generic predictions, naturally led to diagnoses and new non-Keynesian macroeconomic models. But, remarkably, these diagnoses and alternatives were in fact "throwbacks" to classical theory. For at this point the rational expectations theory counter-revolution took hold. It began with a "hypothesis" of John Muth:

> . . . expectations, since they are informed predictions of future events, are essentially the same as predictions of the relevant economic theory. At the risk of confusing this purely

descriptive hypothesis with a pronouncement as to what firms ought to do, we call such expectations 'rational'.

According to at least some rational expectations approaches to the IS-LM curve, governmental action cannot raise nation income by shifting these curves because people are too rational to be tricked by such policies. Or at least enough of them will know that the government embraces Keynesian theory, and will have expectations about the effects of government policies, so that these policies will be ineffective.

For example, the rational expectations theorist argues that when the government sells bonds to cover the deficit it creates (through shifting the IS curve), rational agents realize that servicing this debt will require higher taxes, sooner or later. Accordingly, the alleged wealth effects of holding new government bonds are at least blunted if not completely overborne by the expectations of higher taxes. And this expectation leads to more saving and less consumption, thus lowering national income back in the direction of *Y*. In other words, governmental steps to increase national income by borrowing will not do so if economic agents are rational, and have correct beliefs about the economic model the government employs and its policy-ramifications.

Similarly for the attempt to raise national income through increases in the money supply. The rational agent recognizes that increasing the money supply will reduce the real value of his monetary holdings, by increasing the price-level. Accordingly, he will have to save more money in order to maintain the real value of his money balances. Increased savings means decreased consumption, thereby blunting the original effects of an increase in the money supply.

On this view, neither monetary policy nor fiscal policy can be effective, because it is predicated on the false assumption that economic agents' expectations about the future are not well-informed. Keynesian theory is held to assume that agents' expectations about the future are a function solely of their knowledge about the past: they simply extrapolate current prices and quantities forward into the future. This is misleadingly called the assumption of "adaptive expectations" – misleading because the expectations track past prices, and will not be adaptive if policy changes.

The rational expectations view is not that all economic agents are omnicient about the future, nor even that most agents' expectations are better than merely "adaptive". It holds that at least some agents employ all information available in order to formulate their expectations, not just past values of economic variables; it holds that aggregated expectations about the future are on average correct, or at least that enough agents are correct enough of the time to take economic advantage of governmental policy changes, and that the economic incentives for doing so are great enough for their actions to effect the direction of the entire economy.

The most fervent proponents of rational expectations theory argue that it shows no governmental policies can attain the fine-tuning that Keynesian theory aims at. But this conclusion is far too strong. Even if it were shown that rational expectations will "outsmart" every sort of fiscal and monetary policy based on a Keynesian model, at most this would show that the Keynesian model embraced by the government was faulty, not that no macroeconomic theory of any kind is possible. Rational expectations theory is not a general proof that all macrotheories must fail through "reflexivity" – i.e., as a result of people coming to know about them.

Even among those proponents of rational expectations theory less extreme in their claims about the impotence of macroeconomic policy, the theory is a return to classical orthodoxy, a "reactionary" turn in economic theory. (I do not use the term "reactionary" in a pejorative way here.)

The classical theory is sometimes said to deny the possibility of an unemployment equilibrium because economic agents are rational. Keynes is held to have proved the possibility of an unemployment equilibrium by assuming that people are not fully rational. For example, the stickiness of wages reflects the existence of a "money-illusion" that the classical theory of consumer choice proscribes.

The rational expectations critique of Keynesian theory is that it fails to treat people as economically rational agents, thorough-going optimisers, that it surrenders without warrant the classical assumption that markets all clear at equilibrium, it fails to link up with the well-developed body of price, value, and welfare-economic theory.

As for Keynes' original motivation, to explain the depression by explaining how an equilibrium that didn't clear the labor markets was possible, it simply denies that there is any such thing:

Involuntary unemployment is not a fact or phenomenon which it is the task of theorists to explain. It is, on the contrary a theoretical construct which Keynes introduced in the hope that it would be helpful in discovering a correct explanation for a genuine phenomenon: large-scale fluctuations in measured, total employment. Is it the task of modern theoretical economics to "explain" the theoretical constructs of our predecessors? (Lucas, 1978, pp. 353–7)

And of course, rational expectations theory provides its own account of fluctuations in the business cycle, and the causes of high unemployment. A sketch of one version of the rational expectations theory of fluctuations in employment shows the degree to which argument between rival theories in economics remains at the generic level. For the rational expectations model is provided as at most a *caricature* that renders expectable the direction which phenomena take, though not the dimensions of the phenomena. (The description of such models as 'caricatures' is not mine. It is due to a distinguished economist, Hal Varian, who does not employ the label as a term of abuse, but to highlight something important about economic theory, as we shall see in the next section.)

Lucas (1977) offers the following account of how fluctuations in employment may reflect rational expectations: Consider how a representative agent producing a common good will respond to a change in its price. In the case of a laborer, this good is his work, and the price is his wage. If the price-increase is expected to be permanent, there will be no effect on the level of employment. At any rate data from labor economics suggest this. But if the price (i.e., wage) increase is viewed as temporary, the laborer will increase his supply of labor, he will work more hours now, and postpone his leisure to a period after the wage-rate returns to its prior level. And vice versa when the wage rate is thought to have moved downward temporarily: vacations will be moved forward, since the opportunity cost of leisure has declined. So, labor supply varies with expectations about the duration of wage-rate changes. And employment fluctuates because worker's speculate on the market for leisure – buying more when the price declines, and less when it rises. Small random fluctuations in the wage-rate can cause considerable changes in the employment-level, whence the business cycle, or at least an important part of it.

The reaction of some economists, especially Keynesian, to rational expectations theory is roughly that of not knowing whether to laugh or to cry. They recognize that the theory is an attempt to return to the *status quo ante* Keynes, and surrenders all the hard-won interventionism of the mixed economy. James Tobin is eloquent on this point:

> [Rational expectations theorists] are all inspired by faith that the economy can never be very far from equilibrium. Markets work, excess supplies and demand are eliminated, expectations embody the best available information, people always make any and all deals which would move all parties to preferred positions. With such faith the orthodox economists of the early 1930's could shut their eyes to events they knew a priori could not be happening ... Keynes might say this is where he came in. [Tobin, 1977, p. 461]

The dispute between Keynesians and rational expectations theories is not one for us to adjudicate. But it illustrates quite clearly the fact that two hundred years after Smith and a hundred years after Walras, economic theory is still brought to bear on economic phenomena at most *generically*. One reason may be that economists have still not decided in what direction the long-term facts to be explained point, and so cannot choose which generic theory to attempt to improve in the direction of specificity. Here Leontief's criticism seems to apply. We do not yet have a good enough grip on the quantitative facts to be explained. And too little effort seems to be devoted to determining these facts, as opposed to constructing intellectually beautiful models of possible economies.

On the other hand, there may be an intrinsic limitation on conventional economic theory to generic prediction. So that the rational expectation theorist continued devotion to it simply reflects economic theory's continued commitment to perpetually being at most generic in its claims about the world.

It is evident that rational expectations theorists are not daunted by the "unrealism" of their models, by the fact that in their most attractive explanation for the failure of a "more realistic" theory than the classical one is not just less realistic, but uncompromisingly classical. This is a fact that needs to be explained. One explanation, convenient and widely embraced, is Friedman's: the "unrealism of the assumptions" is no defect in a theory, indeed it is a requirement for real explanatory power. This is an explanation that we can disregard, if only because it rests on a controversial assumption about the predictive success of economic theory. Moreover, it does not *specifically* explain the continuing attachment of economic theory to optimizing models.

Another explanation is that prediction, even generic prediction, is not what successive economic theories aim at. On this view successful generic prediction is at most a necessary requirement on economic models. It's not enough, but it's a start, and the rest of what the economist wants is provided by other entirely non-predictive features of his models.

5. GIBBARD AND VARIAN ON ECONOMIC MODELS AS CARICATURES

It is certainly true that economists are interested in other things along side prediction, specific or generic. Although it is quite wrong to deny that economics has or needs any predictive pretentions, the discipline cannot be focused solely, or even largely on this aim. For if it were, it would long ago have surrendered neoclassical theory for some other more predictively powerful one. These other apparently non-predictive interests are explored, in a paper entitled 'Economic Models', by a distinguished economist, Hal Varian, and an equally distinguished philosopher, Alan Gibbard. What is especially important for our purposes is that in the end, they lead to the conclusion that the best we can hope for from economic models is generic prediction after all, and that this is enough. For there seems no other reason offered to suppose that such models actually provide understanding of economic phenomena.

Gibbard and Varian raise the question, "In what ways can a model help in understanding a situation in the world when its assumptions, as applied to that situation, are false" (Gibbard and Varian, 1978, p. 665).

Its worth restating the point of Section 1 here, that the general question is not at issue. We know how and why theoretical models in physics, for example, help in understanding, even though their assumptions are false. If the same answer – predictive success – is offered for economic models, then the real question, in what way *economic* models can help, will not have been answered.

Gibbard and Varian begin by distinguishing ideal models from descriptive ones which attempt to fully capture economic reality. It is the former which is their concern, evidently because they are both more common, and more important in economic theory. Within this class there is a further distinction between approximations and caricatures – models that purport to give an approximate description of reality and those which "seek to 'give an impression' of some aspect of economic reality not by describing it directly, but rather by emphasizing – even to the point of distorting – certain selected aspects of the economic situation" (p. 665). This distinction, between approximations and caricatures is, they say, one of degree.

Now, Gibbard and Varian note, a model is a story with a specified structure – the story gives something like the extension of some of the predicates in the structure, but it is not a story about any particular producers, or consumers or any economic situation in the world. It may however be applied to a situation. The result is an applied model, produced by giving the structure's predicates and quantifiers a particular extension. Only when thus interpreted, can we ask how close to the truth the model is.

Though false, the model's assumptions are hypothesized by the economist to be close enough approximations to the truth for his explanatory "purposes". This explanation takes the following form:

> First, if the assumptions of the applied model were *true*, the conclusion would be – here the proof is mathematical. Second, the assumption is in fact sufficiently close to the truth to make the conclusions approximately true. For this no argument within the model can be given; it is rather a hypothesis, for and against which evidence can be given. One kind of evidence is evidence for the rough truth of the conclusions of the applied model; another kind is evidence for the rough truth of the assumptions.

The model has explanatory power with respect to its explanandum if we have evidence that "the conclusions of the ... applied model were close to the truth *because* its assumptions were close to the truth" (p. 670).

But closeness to the truth turns out almost always to be a matter of what I have called "generic" prediction or explanation. Models are employed to explain the "central tendency" of economic behavior (p. 670), "when models are applied as approximations, few if any of the degrees of approximation are characterized numerically" (p. 672). Economists, we are told, apply models to situations in two different ways: econometrically and casually. About econometric applications Gibbard and Varian say little, partly because they note there is a well-developed methodology for them, but mainly because such models are usually severe complications of simpler casual models, specifically intended for prediction. As such they do not raise the problems of non-predictive models. On the other hand, we have the testimony of Leontief that such econometric models do not have a good track-record. So, while conceptually unproblematic, they are of arguable utility.

However, what Gibbard and Varian call *casual* application of models, is both more characteristic of economic theory, and far more problematical. Thus, they are right to focus on these applications.

> The goal of casual application is to explain aspects of the world that can be noticed or conjectured without explicit techniques of measurement.... When economic models are used in this way to explain casually observable features of the world, it is important that one be able to grasp the explanation. Simplicity then will be a highly desirable feature of such models (and) complications may be unneccessary, since the aspects of the world the model is used to explain are not precisely measured. (p. 672)

This, notice, is much the same position Samuelson accepted in 1947. Indeed, it is very little different from the explanatory results of Adam Smith's *Wealth of Nations*. Gibbard and Varian expound casual application in ways that make the lack of advance on Smith even more palpable. For they go on to assert that one of the most important explanatory functions of an economic model is to serve as a *caricature*. A model is often designed to exaggerate some feature of reality, instead of approximating it, and this is essential to its explanatory power:

> An applied model that ascribed to (an assumption) its approximate place in reality might bury its effects, and for that reason, a model that is a better approximation to reality may make for a worse explanation of the role of some particular feature of reality.
>
> If the purpose of economic models were simply to approximate reality in a tractable way, then, as techniques for dealing with models are refined and as more complex models become tractable, we should expect a tendency towards a better fit with complex reality through more and more complex models.... [B]ut a tendency to better approximations through more complex models is by no means the rule. (p. 673)

This passage raises many fundamental questions that Gibbard and Varian do not address. For example, what is the cognitive function of explanation, if a more nearly true model may turn out to be a worse explanation. Surely, the psychological accessibility of an explanation is not an important constraint, by comparison to its degree of evidential support, systematical connections, and predictive applicability. Moreover, the fact that the history of economic theory does not reveal a "tendency towards better fit with complex reality" is a matter that cries out for explanation. Gibbard and Varian are right to say that economists prize what they call caricature models. But the reason they do so remains unspecified. One possibility is suggested in the claim that approximating models "bury" their effects. I think this means that such models bring in so many possibly countervailing economic forces that they do not issue even in generic predictions. On this interpretation, caricatures are prized because they do provide at least such generic claims about "general tendencies".

A caricature model is one in which approximating reality is not important, rather it attempts to subsume a complex phenomenon under a simple and relatively one-sided "story". Gibbard and Varian cite an example from Samuelson: a caricature model explaining intergenerational transfers of income by assuming that the population is composed of perfectly rational persons, who live in only two periods: working and retired. They might just as well have sketched one of the rational expectations models, about how people react to macroeconomic policy, that figure in the previous section. About such examples, they ask, when will a caricature model be helpful in understanding a situation? The answer seems to be that a caricature will explain when the explanandum phenomena "*do not* depend on the details of the assumptions," that is, if they are *robust* – i.e., follow from several different caricatures of reality.

But this is deeply perplexing a claim. For the conventional wisdom in the philosophy of science is that a theory has little explanatory power, if the phenomena it purports to explain can also be explained by a wide variety of equally plausible alternative theories. (Economic opponents of rational expectations theory will agree with this claim: consider the failure of Keynesian policies, which many agree is a robust conclusion. But this failure does not in their view tend to substantiate rational expectations assumptions. There are too many alternative possibilities.) Robustness is an important attribute, but presumably for the explanans in an explanation, and not for the explanandum – for in most cases we already know that the explanandum obtains, and we seek to explain it. Of course the robustness of consequences might do credit to a caricature, even in the presence of other equally plausible ones with the same consequence. But only if caricatures were valued for their generic predictions – since this is what having a robust consequence usually comes to among such models. We have some evidence for this interpretation in Gibbard and Varian's statement that in the employment of a caricature, the economists hypothesis is "that a conclusion of the applied model depicts a tendency of the situation, and that this is because the assumptions caricature features of the situation and the conclusion is robust under changes of caricature" (pp. 676–7).

A caricature, we are told, "differs from an approximation ... not only in its simplicity and inaccuracy, but in its deliberate distortion of reality .... to isolate the effects of one of the factors involved in the situation, or to test for the robustness under changes of caricature" (p. 676). These claims are, I concede, descriptively correct ones to make about economic theorizing. But unless there is a connection between robustness of conclusions and the understanding or explanatory power caricatures provide, they still leave open the fundamental question of how "a caricature (which) involves deliberate distortion (can) illuminate an aspect of economic life". Only in Gibbard and Varian's peroration is this issue really broached:

> ... a caricature involves deliberate distortion to illuminate an aspect of economic life. If the uses of deliberate distortion are ignored, and the job of applied models is taken to be no more than accurate approximation under constraints of simplicity and tractability, many of the caricatures economic theorists construct will seem unsuited for their job. (p. 677)

But the issue is not *whether* economists treat caricatures as conveying understanding. Of that there is no doubt. After all, nothing is more evident than the frequency of such models in the literature. What is at issue is *how* such models convey understanding, what kind of understanding is it they convey. What is there that economists are after by way of economic *knowledge*, besides the sort (rather inadequately) provided by approximation models, that makes caricatures so important? These are questions Gibbard and Varian do not specifically address, though their typology of economic models makes them particularly pressing.

6. SUMMARY AND CONCLUSIONS

I have argued not that economics has no predictive content, but that it is limited, or at least has so far been limited to generic predictions. Now this is an important kind of prediction, and almost certainly a necessary preliminary to specific or quantitative predictions. But if the sketch of an important episode in the twentieth century history of the subject I have given is both correct and representative, then economics seems pretty well stuck at the level of generic prediction. And at least some influential economists and philosophers of economics seem well satisfied with stopping at the point of generic prediction. Or at least they give no other reason than its power to produce such predictions as a justification for the character of economic theory. But this leads to the question that is the title of my paper, is generic prediction enough?

REFERENCES

Blaug, M.: 1978, *Economic Theory in Retrospect*, Cambridge University Press, Cambridge.

Friedman, M.: 1953, *Essays in Positive Economics*, University of Chicago Press, Chicago.

Gibbard, A., and H. Varian: 1978, 'Economic Models', *Journal of Philosophy* **75**, 664–77.

Hicks, J.: 1939, *Value and Capital*, Oxford University Press, Oxford.

Leontief, W.: 1951, 'Input-Output Economics', *Scientific American* **185**, 15–21.

Leontief, W.: 1982, 'Academic Economics', *Science* **9**, p. 2.

Leontief, W.: 1985, *Essays in Economics*, Transaction Books, New Brunswick.

Lucas, R. E.: 1977, 'Understanding Business Cycles', in K. Brunner and A. Meltzer (eds.), *Stabilization of the Domestic and International Economy*, North-Holland, Amsterdam.

Lucas, R. E., 1978, 'Unemployment Policy', *American Economic Review* **68**, 353–7.

Muth, J.: 1961, 'Rational Expectations and the Theory of Price Movements', *Econometrica* **29**, 315–35.

Rosenberg, A.: 1976, *Microeconomic Laws: A Philosophical Analysis*, University of Pittsburgh Press, Pittsburgh.

Samuelson, P.: 1947, *Foundations of Economic Analysis*, MIT Press, Cambridge.

Tobin, J.: 1977, 'How Dead is Keynes?', *Economic Inquiry* **15**, 459–68.

Wicksteed, P. H.: 1910, *The Common Sense of Political Economy*, London.

Manuscript received 25 January 1988

Department of Philosophy
University of California
Riverside, CA 92521
U.S.A.

CRISTINA BICCHIERI

SELF-REFUTING THEORIES OF STRATEGIC INTERACTION: A PARADOX OF COMMON KNOWLEDGE*

INTRODUCTION

Game theoretic reasoning is sometimes strikingly inconsistent with observed behavior, or even with evidence from introspection. Famous examples of such inconsistency are the finitely repeated Prisoner's Dilemma game and Selten's Chain Store Paradox (Selten, 1978). In both cases, some plausible solutions run counter to game theoretic reasoning and appear to point to the inadequacy of the game theoretic notion of rationality in capturing important features of human behavior. These considerations do not apply to artificial settings only: in a wide range of ordinary social interactions it does not pay to be (or look) too rational. As Goffman puts it, expert poker players sometimes discover that one can lose the game because of playing too well (Goffman, 1969). In international conflicts, it may well pay to be thought of as a 'mad dog'. And in the above quoted games, all or some of the players involved get a lower payoff than they would were they to play in less than a rational manner. All attempts to explain the emergence of cooperation in a finitely repeated Prisoner's Dilemma, as well as that of 'reputation effects' in the Chain Store story, have required either some version of 'imperfect rationality' (Selten, 1978) or a change in the structure of the game, such as assuming altruism (and thus changing the payoffs), or imposing incomplete information at the beginning of the game (Kreps et al., 1982).

The attempts at bridging the gap between the correct but implausible game theoretic results and the plausible (but unexplained) observed outcomes have assumed the logical inescapability of the classic game theoretic solution. However, as much as it is true that only under certain conditions can one behave as a 'mad dog', or in less than a rational manner, it is also the case that the standard game theoretic solution can only obtain under special conditions. These conditions, as I shall argue, pertain to the players' knowledge of the theory of the game (i.e., the theory of backward induction in finite games of perfect information). Here by 'theory' I mean the following

Erkenntnis **30** (1989) 69–85.

things: a set of assumptions about the players' rationality and their beliefs about each other's rationality, a specification of the structure of the game, of the players' strategies and payoffs and the hypothesis that structure, strategies and payoffs are known by the players. From these assumptions the unique equilibrium solution is derived.

How much do the players need to know about the game for them to successfully complete the reasoning required of them, and infer the unique solution? Intuitively, one might expect that the more the players know about each other, the easier it should be for them to replicate each other's reasoning and to predict each other's play. To prove otherwise is the aim of the present paper. In fact, it can be shown that in order for the backward induction solution to obtain, the players must have some knowledge of the theory's assumptions, but no common knowledge of them.[1] The paradoxical conclusion we reach is that *common knowledge of the theory of the game makes the theory inconsistent.*

An obvious requirement a theory of the game has to satisfy is that it be free from contradictions at every information set (Reny, 1987). If a player were to find herself at an information set with which the theory of the game she is using is inconsistent, she would be deprived of a theory upon which to base her decisions. This would leave the other players (and the game theorist) without a theory, too, since they would become unable to predict what she will do, and would therefore be unable to decide what to do themselves. Consistency and predictability are strictly related. In perfect information games, it is enough to assume the players to have common knowledge of the assumptions regarding their beliefs to make the assumptions inconsistent with some information set. This conclusion implies that once common knowledge of the theory of the game is assumed, it is no longer necessary to change the structure of the game, or to reject the traditional definition of rationality, to allow for different solutions. Indeed, the existence of common knowledge makes deviations from the classical solution plausible, and compatible with individual rationality.

Knowing the disruptive effects of common knowledge of beliefs over the theory of the game, the players may have an incentive to manipulate knowledge in order to attain higher payoffs. Manipulation of knowledge, in this context, means that one or more players would communicate their beliefs to the rest of the players. At the end of the paper, I shall only briefly suggest how this result can bear upon the cooperative solution in the finitely repeated Prisoner's Dilemma and the 'reputation effects' of the Chain Store Paradox.

BACKWARD INDUCTION

The games I am going to discuss are finite two-person extensive form non-cooperative games of perfect information. A non-cooperative game is a game in which no precommitments or binding agreements are possible. By 'extensive form' is meant a description of the game indicating the choices available to each player in sequence, the information a player has when it is his turn to move, and the payoffs each player receives at the end of the game. Perfect information means that there are no simultaneous moves, and that at each point in the game it is known which choices have previously been made. According to the backward induction theory (Kuhn, 1953), any such game has a unique solution. Take as an example the following game:

G_1

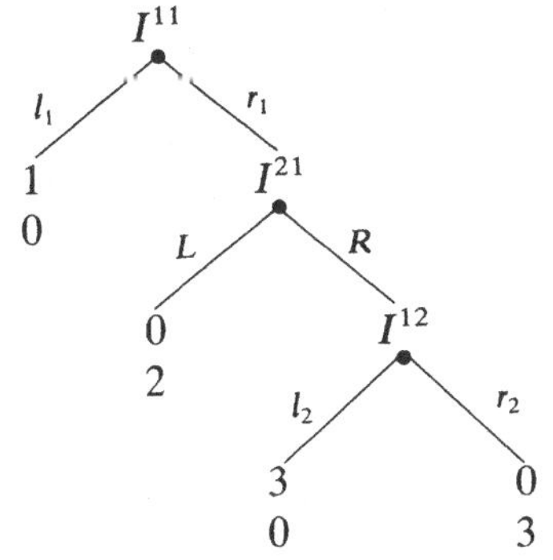

I^{ij} denotes the j-th information set ($j \geq 1$) of player $i (i = 1, 2)$. Since there is perfect information, I^{ij} is a singleton set for every i and j. Each player has two pure strategies: either to play 'left', thus ending the game, or to play 'right', and allow the other to make a choice. The game starts with player 1 moving first. The payoffs to the players are represented at the endpoints of the tree, the upper number being the payoff of player 1, and each player is assumed to wish to maximize his expected payoff. The game is played sequentially, and at each node it is known which choices have been previously made. Player 1, at his first node, has two possible choices: to play l_1 or to play r_1. What he chooses depends on what he expects player 2 to do afterwards. If he expects player 2 to play L at the second node with a high probability, then it is rational for him to play l_1 at the first node; otherwise he plays r_1. His conjecture about player 2's choice at the second node is based on what he thinks player 2 believes would happen if she played R. Player 2, in turn, has to conjecture what player 1 would do at the third node, given that she played R. Indeed, both players have to conjecture each other's beliefs and conjectures at each possible node, until the end of the game.

The classical solution of such games is obtained by backward induction as follows: at node I^{12} player 1, if rational, will play l_2, which grants him a maximum payoff of 3. Note that player 1 does not need to assume 2's rationality in order to make his choice, since what happened before the last node is irrelevant to his decision. Thus node I^{12} can be substituted by the payoff pair $(3, 0)$. At I^{21} player 2, if rational, will only need to believe that 1 is rational in order to choose L. That is, player 2 need consider only what she expects to happen at subsequent nodes (i.e., the last node) as, again, that part of the tree coming before is now strategically irrelevant. The penultimate node can thus be substituted by the payoff pair $(0, 2)$. At node I^{11}, rational player 1, in order to choose l_1, will have to believe that 2 is rational *and* that 2 believes that 1 is rational (otherwise, he would not be sure that at I^{21} player 2 will play L). From right to left, nonoptimal actions are successively deleted (the optimal choice at each node is indicated by doubling the arrow), and the conclusion is that player 1 should play l_1 at his first node.

In the classical account of such a game, this represents the only possible pattern of play by rational players. Note, again, that specification of the solution requires a description of what both agents expect to happen at each node, were it to be reached, even though in equilibrium play no node after the first is ever reached. Thus the solution concept requires the players to engage in hypothetical reasoning regarding behavior at each possible node, even if that node would never be reached by a player playing according to the solution.

BELIEFS, ITERATED BELIEFS, COMMON KNOWLEDGE

The theory of the game we have just described makes a series of assumptions about players' rationality, knowledge and beliefs, from which the backward induction (b.i.) solution necessarily follows. Let us consider them in turn. First of all, the players have to have k-th level knowledge of their respective strategies and payoffs (if the game has $k+1$ stages). Second, the players must be rational, in the sense of being expected utility maximizers. Third, the players are assumed to believe each other to be rational and, depending on the length of the game, to have iterated beliefs of k-th degree about each other's rationality.

One property generally required of an agent's beliefs is that they be internally consistent. Thus, for example, player i cannot believe that j is rational and not expect j to choose his best response strategy. It must be added that in game theory the notions of knowledge and belief are state-based, where the state a player is at, is his information set. An agent i cannot possibly believe p at information set I^{ij} if his being at that information set contradicts p. For the purposes of our discussion, we require an individual's beliefs to have two properties: (a) they must be internally consistent, and (b) i's beliefs at any information set must be consistent with the information available to the player at that information set.

The language in which we are going to express game-theoretic reasoning is a propositional modal logic for m agents. Starting with primitive propositions p, q, ..., more complicated formulas are formed by closing the language under negation, conjunction, and the modal operators $B_1, \ldots, B_m$ and $K_1, \ldots, K_m$ (Hintikka, 1962). The very idea of iterated beliefs, however, requires a generalization of the notion of belief from an individual i to a group G (Halpern and Moses, 1986). Let us define $\mathbf{B}_G p$ ('everyone in G believes p') in the following way: $\mathbf{B}_G p$ holds iff all members of G believe p. Formally,

$$\mathbf{B}_G p \equiv \bigwedge_{i \subset G} B_i p$$

$\mathbf{B}^k_G p$, $k \geqslant 2$ ('p is $\mathbf{B}^k$-belief in G') is defined by

$$\mathbf{B}^1_G p = \mathbf{B}_G p,$$
$$\mathbf{B}^{k+1}_G p = \mathbf{B}_G \mathbf{B}^k_G p, \text{ for } k \geqslant 1$$

p is said to be $\mathbf{B}^k$-belief in G if "everyone in G believes that everyone in G believes that... that everyone in G believes that p is the case" holds, where the phrase "everyone in G believes that" appears in the sentence k times. Equivalently,

$$\mathbf{B}^k_G p \equiv \bigwedge_{i_j \in G,\, 1 \leq j \leq k} B_{i_1} B_{i_2} \ldots B_{i_k} p$$

There are circumstances in which we may require the agents' beliefs to be common knowledge. For example, if it is publicly announced that "I believe that you believe that I am rational", everybody will know that I believe that you believe that I am rational, and everybody will know that everybody knows that . . . , and so on ad infinitum. If $K_i p$ stands for "i knows p", let us define "everybody knows p" as

$$\mathbf{K}_G p \equiv \bigwedge_{i \in G} K_i p$$

Iterated knowledge of p can be thus expressed:

$$\mathbf{K}^k_G p \equiv \bigwedge_{i_j \in G,\, 1 \leq j \leq k} K_{i_1} K_{i_2} \ldots K_{i_k} p$$

Let us now define $\mathbf{C}_G p$('p is common knowledge in G') as follows: p is said to be common knowledge in G if p is true, and is $\mathbf{K}^k_G$-knowledge for all $k \geq 1$. In other words,

$$\mathbf{C}_G p \equiv p \wedge \mathbf{K}_G p \wedge \mathbf{K}^2_G p \wedge \ldots \wedge \mathbf{K}^m_G p \wedge \ldots.$$

In particular, $\mathbf{C}_G p$ implies all formulas of the form $K_{i_1} K_{i_2} \ldots K_{i_n} p$, where the i_j are all members of G, for any finite n, and is equivalent to the infinite conjunction of all such formulas. Clearly, the notions of group knowledge introduced above form a hierarchy, with $\mathbf{C}_G p \supset \cdots \supset \mathbf{K}^{k+1}_G p \supset \cdots \supset \mathbf{K}_G p \supset p$.[2] A similar hierarchy is formed by the notions of group beliefs, e.g., $\mathbf{B}^{k+1}_G p \supset \cdots \supset \mathbf{B}_G p$.

It is easy to verify that in game G_1 (as in any game of perfect information) every two levels of the belief hierarchy can be separated, in that there will be an action for which one level in the hierarchy will suffice, but no lower level will. At different stages of the game, one needs different levels of beliefs for backward induction to work. For example, if R_1 stands for 'player 1 is rational', R_2 for 'player 2 is rational', and $B_2 R_1$ for 'player 2 believes that player 1 is rational', R_1 alone will be sufficient to predict 1's choice at the last node, but in order to predict 2's choice at the penultimate node, one must know that rational player 2 believes that 1 is rational, i.e., B_2R_1. B_2R_1, in turn, is not sufficient to predict 1's choice at the first node, since 1 will also have to believe that 2 believes that he is rational. That is, $B_1B_2R_1$ needs to obtain. Moreover, while R_2 only (in combination with B_2R_1) is needed to predict L at the penultimate node, B_1R_2 must be the case at I^{11}. More generally, for an N-stage game, the first player to move will have to have the $N-1$-level belief that the second player believes that he is rational ... for the b.i. solution to obtain.

DISTRIBUTED KNOWLEDGE AND FULL KNOWLEDGE

It has been argued that at I^{21} it is by no means evident that player 2 will only consider what comes next in the game (Binmore, 1987; Reny, 1987). Reaching I^{21} may not be compatible with a theory of backward induction, in the sense of not being consistent with the above stated assumptions about players' beliefs and rationality. Indeed, I^{21} *can only be reached if 1 deviates from his equilibrium strategy*, and this deviation stands in need of explanation. When player 1 considers what player 2 would choose at I^{21}, he has to have an opinion as to what sort of explanation 2 is likely to find for being called to decide, since 2's subsequent action will depend upon it. Obviously enough, different explanations lead to different expected payoffs from making the same move leading to I^{12}.

What player 2 infers from 1's move, though, depends on what she believes about player 1. Up to now, we know that different players need different levels of beliefs for the b.i. solution to obtain. More precisely, the theory of the game assumes that the players make use of all of the propositions in '$R_1 \wedge R_2 \wedge B_2R_1$' (which stands for '1 is rational and 2 is rational and 2 believes that 1 is rational'). It might be asked whether it makes a difference to the backward induction solution that the theory's assumptions about players' beliefs are known to the players. This might mean several things. On the one hand, the theory's assumptions can be 'distributed' among the players, so that not all players have the same information. That is, beliefs attributed to the players by the theory are differentially distributed among them, as opposed to the case in which all players share the same beliefs. In this latter case, all players are endowed with the same information. In both cases, the players do not know what the other player believes.

We may imagine the players being two identical reasoning machines programmed to calculate their best action which are 'fed' information in the form of beliefs. The machines are capable of performing inferences based upon the available information, which consists of 'beliefs' about the other machine. A machine can be fed more, less, or the same information as another machine. Let us look first at the case in which the beliefs 'fed' to each machine are the minimal set consistent with successful backward induction. Each player can infer about the other what his own beliefs allow her to, and no more. In fact, this allocation of beliefs is implicit in the classical solution.[3] Assuming the players to be rational, beliefs are thus distributed:

Player 1 believes:	Player 2 believes:
R_2	R_1
B_2R_1	

Evidently, 2 ***does not know*** that 1 believes R_2, nor that 1 believes that she believes R_1. But since she believes R_1, she plays L at I^{21}. Given her belief, the only inference that 2 can draw from being at I^{21} is that player 1 chose r_1 either because he does not believe that player 2 is rational (i.e., $\sim B_1R_2$), or does not believe that 2 believes that he is rational (i.e., $\sim B_1B_2R_1$), or any combination thereof. Thus player 2's knowledge of the game and beliefs allow the play of r_1 by rational player 1, since her belief that 1 is rational is not contradicted by reaching mode I^{21} (I assume that if a belief is consistent with reaching an information set, then that belief is maintained). It follows that 2's rational response is still L. Player 1 does not know what 2 believes, but he believes R_2 and B_2R_1; therefore he should play l_1, whereas 2 does not know that he should choose it. It must be noticed that this conclusion follows both from players' rationality and from distributed knowledge of beliefs (and iterated beliefs) among them.

It is easy to verify that, were the players to have the same beliefs, the backward induction solution would still obtain. In this case, everybody has to believe that '$R_1 \wedge R_2 \wedge B_2R_1$' is true. Given that it is redundant to have a player believe that he or she is rational or believes something (they are supposed to be rational and to know what beliefs they have), this distribution of knowledge in no way modifies the conclusion that a deviation from equilibrium is consistent with the players' beliefs (and therefore with the assumptions of the theory).[4] Thus backward induction works even if all players know the same set of propositions. In both full and distributed knowledge, however, the players have been assumed not to know what the other believes.

COMMON KNOWLEDGE

Intuitively, one might expect that the more the players know about the theory of the game, the more enhanced their (and the theory's) predictive capability would be. That is, the more the players know about each other's knowledge and beliefs, the more they become able to fully replicate the opponent's reasoning. In what follows, it will be assumed that the players have common knowledge of the theory's assumptions regarding their beliefs. That is, all players know that all players believe that '$R_1 \wedge R_2 \wedge B_2R_1$' is true, and they all know that they all know, ... ad infinitum. The paradoxical conclusion is that a theory that is common knowledge among the players becomes inconsistent.[5] To see why common knowledge of beliefs leads to inconsistency, let us detail what each player knows under this condition:[6]

Player 1 knows:	Player 2 knows:
B_2R_2	B_1R_1
B_2R_1	B_1R_2
$B_2B_1R_2$	$B_1B_2R_1$
⋮	⋮

To get the backward induction solution, such an infinite chain of beliefs is not even necessary. The players need only both believe that '$R_1 \wedge R_2 \wedge B_2R_1$' is true. Thus player 1 should choose l_1 at node I^{11}. Suppose that I^{21} were reached. Player 2 knows $B_1R_1 \wedge B_1R_2 \wedge B_1B_2R_1$. But, since the node has been reached, one or more of the conjunction's elements must be false. Common knowledge of beliefs does not allow player 2 to assume that either $\sim B_1B_2R_1$ or $\sim B_1R_2$ is the case, and this is common knowledge. The deviation can only be explained assuming that $\sim R_1$; in this case, 2 would respond to r_1 with R.

But can $\sim R_1$ be assumed? Both players are rational; each knows he is rational, but does not know that the other is rational. So much is postulated by the theory of the game. If common knowledge of beliefs is the case, each player will know that the other believes himself rational. Whereas one cannot be rational without knowing it (there is no such thing as 'unconscious' rationality), does knowing that somebody believes himself rational mean knowing that he is in fact rational? In general, the fact that somebody believes that p in no way implies that that person knows p, for one may know only true things, but believe many falsehoods. If p were false, one could not know that p, but still believe that p is the case.

Yet the implicit and explicit assumptions that game theory makes about the players allow one to infer from i's belief that he is rational that i knows that he is rational. Let us consider them in turn: (i) throughout game theory, it is implicitly assumed that the meaning of rationality is common knowledge among the players. The players know that being rational means maximizing expected utility, and know that they know,... Were a player to use another rule, he would know he is not rational (as one cannot be 'unconsciously' rational, one cannot be 'unconsciously' not rational). *A fortiori*, he could never believe he is rational. Still, it is possible that a player is rational but lacks the calculating capabilities required to compute the equilibrium solution (or solutions), or has a mistaken perception of his payoffs and strategies. In this case, knowing that a player is rational is not sufficient to predict his moves. We thus need to add the following clauses: (ii) the players are perfectly able to follow through the reasoning process, as complicated as it may be, and (iii) the players have common knowledge of the complete description of the game. This means each player knows his (and the other's) payoffs and strategies, and knows that the other knows,... And this rules out misperception.

If common knowledge of their respective beliefs thus implies common knowledge of rationality, it follows that $\sim R_1$ cannot be assumed. But then, of course, we know that 2 cannot assume 1 not to believe R_2, nor can she believe that 1 does not believe B_2R_1. If rationality is common knowledge, the conjunction $R_1 \wedge R_2 \wedge B_2R_1$ must be true, but then a deviation from equilibrium is inconsistent with rationality common knowledge. In other words, a deviation from equilibrium leads player 2 to uphold the following pair of inconsistent beliefs at node I^{21}: $B_2R_1 \wedge B_2(R_1 \rightarrow \sim B_2R_1)$. If the second belief is true, it is not possible that 2 believes 1 to be rational, since that very belief implies that 1 is not rational, contrary to what 2 believes. Assuming common knowledge of the theory's assumptions about players' beliefs would thus render the theory inconsistent at node I^{21}.

If a deviation from equilibrium occurs, is it really the case that 1 is not rational, or is he only trying to cheat player 2 into responding with R? Both things are possible, and there is just no way for player 2 to rationally decide in favor of one of them. Given that this reasoning process is virtual, in the sense that it takes place in the mind of each player before the game starts, player 1 will be unable to predict what 2 would choose were he to deviate from the b.i. solution, since his very deviation would make the assumption that both believe '$R_1 \wedge R_2 \wedge B_2R_1$' to be true inconsistent with reaching node I^{21}.

Indeed, *allowing common knowledge of the theory of the game makes that theory inconsistent*. Thus, were the players to tell each other what they believe, or were a public announcement made stating that they both believe $R_1 \wedge R_2 \wedge B_2R_1$, the theory of the game would immediately become inconsistent.[7]

MANIPULABILITY OF KNOWLEDGE

These findings suggest that the players, knowing this 'common knowledge effect', may have an incentive to use knowledge strategically (i.e., they may communicate their respective beliefs to each other). If the players' beliefs were assumed to be common knowledge in game $\mathbf{G}_1$, for example, it would be common knowledge that player 2 would not know how to interpret a deviation on the part of 1. It would be common knowledge that the theory of the game becomes inconsistent upon reaching the second node, hence player 1 might have an incentive to deviate from the backward induction solution. Since the *choice* of deviating becomes indistinguishable from irrational behavior, alternative solutions are made possible. By 'irrational' behavior I mean the choice of actions which do not maximize expected utility. Irrational behavior does not correspond with erratic behavior; if it were so, one would become unable to distinguish, in a finite number of repetitions, rational from irrational behavior, since rational and random choices might happen to coincide. This fact, however, would deprive the very idea of rationality of its predictive power, since a sequence of choices that appear to be rational might just be the result of chance. In what follows, irrational behavior is understood to mean automatically playing 'right' at every node, irrespective of one's beliefs or knowledge.[8]

Suppose now that common knowledge of beliefs is not assumed (as is the case with backward induction theory). *Would the players have an incentive to make their beliefs common knowledge*? If so, how can we incorporate this new strategic element into the structure of the game? For common knowledge of beliefs to obtain, the players have to communicate to each other their respective beliefs. Since we want this communication to be part of the game, we may think of 'belief communication' as a type of action, and thus add to the original game two possible choices for each player: to communicate beliefs (CB) or not to communicate beliefs (~CB).

The players start the game with the distributed knowledge of beliefs assumed by backward induction theory (e.g., player 1 believes the conjunction $R_1 \wedge R_2 \wedge B_2R_1$ to be true, and player 2 believes the conjunction $R_2 \wedge R_1$ to be true). Adding the choices of communicating or not communicating their beliefs changes the original game $\mathbf{G}_1$ into the following game:

$\mathbf{G}_2$

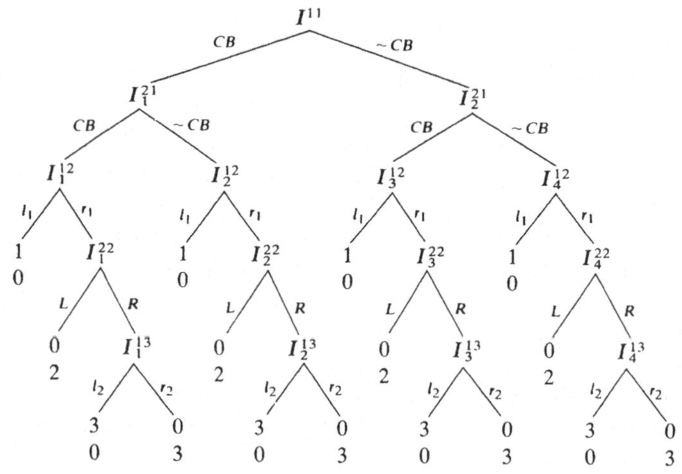

I_p^{ij} denotes the j_p-th node $(j \geqslant 1, p \geqslant 0)$ of player $i (i = 1, 2)$. The game starts with player 1 moving first: he can choose to communicate his beliefs (CB) or not to communicate them (~CB). Player 2 moves afterwards, and she, too, has the choice of communicating or not communicating her beliefs. What is communicated is a statement to the effect that a player believes something to be the case. Let us call this statement **p**; in order for **p** to become common knowledge, however, **p** must be true (otherwise what becomes common knowledge is that a statement **p** has been made, not its truth value). Thus for **p** to be common knowledge it must be evident (and common knowledge)

to both players that *player 1 would not have an incentive to communicate* **p** *if* **p** *were not true*.

Suppose player 1 says that he believes the conjunction $R_1 \wedge R_2 \wedge B_2R_1$ to be true. Let **p** stand for the statement "I believe '$R_1 \wedge R_2 \wedge B_2R_1$' to be true". How is player 2 going to decide that **p** is true? If **p** were false, at least one of the conjuncts would be false, and the question then is whether player 1 would have any incentive to communicate **p** nonetheless. There are three possible cases:

(i) $\sim R_1$ is the case. An irrational player has no incentive to communicate that he is rational, since in any case he is going to deviate from his equilibrium strategy; indeed, even if $\sim R_1$ were false, player 1 would have an incentive to communicate that he does not believe he is rational, so as to 'cheat' player 2 into responding with R. Hence if B_1R_1 is communicated, it must be true:[9]

(ii) $\sim B_1B_2R_1$ is the case. Communicating it, together with B_1R_1, would make player 2 change her belief, were she to believe 1 to be irrational, since B_1R_1 must be true. But then a deviation from the equilibrium strategy would make the theory of the game (from 2's viewpoint) inconsistent with reaching node I_1^{21}. Hence if $B_1B_2R_1$ is told, it must be true;

(iii) $\sim B_1R_2$ is the case. Unless player 2 were to communicate that she is rational, rational player 1 would have a reason to deviate from the equilibrium strategy and no reason to tell a lie. Telling B_1R_2 and believing otherwise would not change his strategy, nor would it change player 2's strategy if 2 is irrational as 1 believes. Then if B_1R_2 is communicated it must be true.

However, the fact that one does not have an incentive to tell a lie does not make what one says necessarily true. *Communicating does not necessarily make* **p** *common knowledge*. Common knowledge of **p** is *supported* by a set of consistent beliefs that the players *might* entertain, but these beliefs are not the only possible consistent beliefs about what player 1 would communicate. That is, if player 2 were to argue along the lines depicted in points (i)–(iii), she would believe **p** to be true, and conversely, if player 1 were to believe 2 thus believes, he would believe that 2 believes that **p** is true. It is therefore possible for 1 to communicate **p** and to believe that 2 holds beliefs that make **p** common knowledge. Even if common knowledge of **p** does not necessarily follow from communicating **p**, *knowing that there exists a consistent set of beliefs supporting common knowledge of* **p** *is enough to induce player 1 to deviate from his equilibrium strategy.*

Does player 1 have an incentive to keep silent? The answer is negative. Since at the start of the game the players are endowed with distributed knowledge of beliefs, player 1 *does not know* how player 2 is going to interpret his silence. Given his beliefs about player 2 (i.e., that 2 is rational and believes 1 to be rational) 1 can expect 2 to respond to a deviation with *L*. Hence it is always better for 1 to communicate **p**, but player 2 does not know that since, by assumption, 2 does not know 1's beliefs at the start of the game.

What about player 2? It is easy to verify that, whatever 1 does, 2 can either keep silent or communicate her beliefs. If **p** is communicated and 2 keeps silent, a deviation from equilibrium can occur, but a deviation can occur if 2 communicates her beliefs, too. It might be thought that, were 1 to keep silent, player 2 would have an incentive to communicate a false belief: that is, that she is not rational. In this case, though, both players would know 2 has an incentive to tell a false belief, so that 2's purpose is defeated.

The idea that knowledge can be thus manipulated has interesting applications. In the finitely repeated Prisoner's Dilemma, it is well known that cooperation can result when the players' rationality is not common knowledge among them (Kreps et al., 1982). In this case both players have a motive to deviate from their classical equilibrium strategies, since the expected payoff of, say, a tit-for-tat strategy is greater than what can be obtained by using the b.i. solution. Instead of assuming incomplete knowledge of each other's rationality on the part of the players, the solution proposed by Kreps et al. can be rationalized in terms of a larger game in which the players have the choice to communicate their beliefs. If they do, this makes it possible for both to deviate from the non-cooperative equilibrium. The same considerations apply to Selten's Chain Store Paradox (Selten, 1978), with the difference that only the Chain Store benefits from deviating from the classical solution. If so, the Chain Store should rationally choose to communicate its beliefs, so as to make the theory of the game inconsistent and to prevent the competitors from using it.

The point here is not that of *predicting* cooperation or reputation effects. They may or may not occur, depending on such elements as the players' psychological propensities, their previous histories and experience, and their capability to interpret the other player's moves. The relevant consideration is that *alternative solutions can be shown to be fully compatible with the players' rationality*. For example, in a Prisoners' Dilemma game repeated 100 times, player 1 may decide to cooperate (C) in the first round, and for the next rounds $N = 2, \ldots, T < 100$ to choose C in period N unless player 2 chose to defect (D) in period $N-1$. For rounds $N > T$, he will always defect, regardless of the other player's choice. Were 2 to play D in period $N-1$, 1 will respond with D in period N. He may keep playing D until player 2 chooses C, and then play C again. However, he may signal to player 2 his willingness to cooperate by returning to play C immediately after he played D in the previous round. Or they may alternate in playing C and D. In general, it can be shown that a cooperative pattern is better for both, and that there are several cooperative equilibria. The precise solution, however, is impossible to predict, both because there are many possible patterns of cooperation, and because each player will probably make a different 'guess' as to the magnitude of T. Indeed, each cooperative equilibrium assumes the players to have common knowledge of their possible strategies, and of the probabilities each assigns to the other's strategies.

In the Chain Store Paradox, backward induction dictates that the chain store play cooperatively with every competitor, and that each competitor enter the local market. While in the short run the cooperative response is more advantageous, we know that in the long run the aggressive response may be a better choice, since it would discourage possible competitors from entering the market. If the game has N periods and there are N competitors, one for each period, the chain store may decide to play aggressive (A) for $N-T$ periods in response to a competitor entering the market, and to cooperate (C) in the remaining T periods. The pattern of play is, however, unpredictable. It depends on such elements as how successful the threat is, and on players' expectations as to the size of T. In this case, too, the 'aggressive' solutions depend on the players having common knowledge of the strategies and probabilities.

In both cases, alternative solutions can be rationally justified without having to introduce notions such as bounded rationality, altruism, or incomplete information. Provided the players are able to evaluate the strategic effect of introducing common knowledge of the theory of the game (through common knowledge of beliefs), they can plausibly decide to communicate their beliefs. This move will open up different patterns of play that provide them with higher expected utilities.

NOTES

* I am grateful to Tommy Tan and Philip Reny for helping me appreciate the importance of common knowledge in games, and to Jon Elster and Michael Woodford for many useful comments. Financial support from National Science Foundation Grant SES 87–10209 is gratefully acknowledged.

[1] For the players to have *common knowledge* that p means that, not only does everyone know that p is true, but everyone knows that everyone knows, everyone knows that everyone knows that everyone knows, and so on ad infinitum.

[2] $\mathbf{K}_G p \supset p$ because $Kp \supset p$ (i.e., one cannot know something which is not true).

[3] This point is seldom recognized by game theorists. Reny discusses the importance of players' knowledge of the theory's assumptions for the b.i. solution to obtain; however, he seems to assume that without common knowledge of backward induction "a more precise treatment of how it operates is called for", since "if one of the players is not familiar with backward induction logic, then he may not play according to its prescriptions. In this case other players (even those familiar with backward induction) may rationally choose not to play according to the prescriptions of backward induction" (Reny 1987, p. 48). As I show, common knowledge of the theory of the game is neither necessary nor sufficient to obtain the b.i. solution.

[4] It is easy to verify that both sentences "all players believe $R_1 \wedge R_2 \wedge B_2R_1$" and "player 1 believes $R_2 \wedge B_2R_1$ and player 2 believes R_1" translate into $\mathbf{B}_G R_2 \wedge \mathbf{B}_G R_1 \wedge \mathbf{B}^2_G R_1$ (where $G = 1, 2$).

[5] Phil Reny has independently shown that assuming the players to have common knowledge of their respective rationality makes the theory inconsistent at some information set (Reny, 1987). I obtain the same result assuming that the players have commmon knowledge of the theory's hypotheses regarding their beliefs. From this assumption, common knowledge of rationality naturally follows.

[6] If we substitute the sentence "all players believe the statement '$R_1 \wedge R_2 \wedge B_2R_1$' to be true" with p, common knowledge that p corresponds to the infinite conjunction: $p \wedge K_1p \wedge K_2p \wedge K_1K_2p \wedge K_2K_1p \wedge \ldots$.

[7] I have shown elsewhere that a richer theory of the game (i.e., a theory that includes a model of belief revision) can accommodate common knowledge of rationality without giving rise to inconsistencies (Bicchieri, 1988a, 1988b).

[8] There are of course many alternative possible interpretations of irrational behavior. For simplicity, I assume the players to have common knowledge that they can be either rational or irrational in the special sense specified above.

[9] In the previous section, it has been shown that if a player believes he is rational, he also knows he is rational, thus if player i says B_iR_i it means i knows he is rational and therefore he is rational.

REFERENCES

Aumann, R. J.: 1976, 'Agreeing to Disagree', *The Annals of Statistics* **4**, 1236–9.
Bicchieri, C.: 1988a, 'Strategic Behavior and Counterfactuals', *Synthese* **75**, 1–35.
Bicchieri, C.: 1988b, 'Common Knowledge and Backward Induction: A Solution to the Paradox', in M. Vardi (ed.), *Theoretical Aspects of Reasoning about Knowledge*, Morgan Kaufman Publishers, Los Altos.
Binmore, K.: 1987, 'Modeling Rational Players', Part I, *Economics and Philosophy* **3**, 179–214.
Goffman, E.: 1969, *Strategic Interaction*, University of Pennsylvania Press, Philadelphia.
Halpern, J. and Y. Moses: 1986, 'Knowledge and Common Knowledge in a Distributed Environment', *Research Report*, IBM Almaden Research Center.
Hintikka, J.: 1962, *Knowledge and Belief*, Cornell University Press, Cornell.
Kreps, D., P. Milgrom, J. Roberts and R. Wilson: 1982, 'Rational Cooperation in the Repeated Prisoner's Dilemma', *Journal of Economic Theory* **27**, 245–52.
Kuhn, H. W.: 1953, 'Extensive Games and the Problem of Information', in H. W. Kuhn and A. W. Tucker (eds.), *Contributions to the Theory of Games*, Princeton University Press, Princeton.
Lewis, D.: 1969, *Conventions*, Harvard University Press, Cambridge.
Luce, R. and H. Raiffa: 1957, *Games and Decisions*, Wiley, New York.
Reny, P.: 1987, 'Rationality, Common Knowledge, and the Theory of Games', *mimeo*, Department of Economics, University of Western Ontario.
Rosenthal, R.: 1981, 'Games of Perfect Information, Predatory Pricing and the Chain-Store Paradox', *Journal of Ec. Th.* **25**, 92–100.
Schelling, T.: 1960, *The Strategy of Conflict*, Oxford University Press, New York.
Selten, R.: 1978, 'The Chain-Store Paradox', *Theory and Decision* **9**, 127–59.
Tan, T. and S. Werlang: 1986, 'On Aumann's Notion of Common Knowledge – An Alternative Approach', *Working Paper* 82–26, University of Chicago.

Manuscript submitted 25 January 1988
Final version received 31 March 1988

The University of Chicago
Center for Ethics, Rationality and Society
5828 South University Avenue
Chicago, IL 60637
U.S.A.

Department of Philosophy
University of Notre Dame
Notre Dame, IN 46556
U.S.A.

ADOLFO GARCÍA DE LA SIENRA*

OPEN PROBLEMS IN THE FOUNDATIONS OF PRICE FORMATION DYNAMICS

ABSTRACT. The aim of the present paper is to attack some of the conceptual problems that arise when the framework of mathematical learning theory is applied to the description of the behavior of the firm, in setting prices and production quotas, in a competitive market. The goal is to depict the process by which the firm fixes prices and production quotas as a stochastic learning process. A solution to such problems is proposed which is based on statistical-decision concepts. The conceptualization of the behavior of the firm by means of concepts pertaining to mathematical learning theory gives rise to certain mathematical problems, which are formulated here in rather precise terms.

It is usual to find in competitive markets that one and the same type of good (perhaps with some slight variations in quality, presentation and so on) is produced by different firms. A typical example nowadays is the proliferation in the markets of North America of the so-called PC-compatible personal computers. There are certain lines among these machines which are so similar, that for all practical purposes they can be considered as equal. When a number of firms are producing the same type of good under competitive conditions, the process of price formation has certain peculiarities which are not present under monopolistic conditions. The aim of price formation dynamics (PFD) is precisely to provide a theory of the process of price formation under competitive conditions (although oligopoly is one particular, limit case) for firms producing the same type of good. The usual story is that each firm has a closed and convex production set Y representing its limited technological knowledge, that it faces a price system $\mathbf{p} = \langle p_1, \ldots, p_n \rangle$ (where n is the number of goods in the economy), and that its behavior consists in choosing a point $\mathbf{x}^*$ in Y that maximizes its profits given the price $\mathbf{p}$, subject to restrictions in initial capital availability and credit. The real story, however, might be different and less deterministic. The contention of PFD is that it is the firm of the agent that in fact fixes both the price and the production quota for its goods and, moreover, that the process of fixing these magnitudes is one of trial and error. More precisely, the claim of PFD is that the process by which the firm fixes prices and production quotas

Erkenntnis **30** (1989) 87–99.

can be adequately modelled within mathematical learning theory (MLT) as a stochastic learning process. What this means is that PFD is an economic interpretation of mathematical learning theory. *If this interpretation is adequate, then we can expect to reduce the problem of the existence of price equilibria to the problem of computing the probability of such equilibria under certain conditions.* To clarify in general these conditions is one of the tasks of the theory.

Some results have been already obtained in the field of PFD. The discrete case of oligopoly and duopoly has been dealt with by Suppes and Atkinson (1960). The results of an experiment designed to test this duopoly theory were reported in Suppes and Carlsmith (1962), where the authors pointed out that the mathematical problems involved in the construction of the continuous analogue of the theory "are substantial, although not insuperable". As we shall see, these problems are indeed substantial, but the very attempt to *conceptualize* in terms of MLT the behavior of the firms, when they must choose a price and a production quota from a continuum of responses, gives rise to no less serious difficulties. My aim in the present paper is to present the main conceptual open problems that have to be solved in order to construct the continuous case of duopoly (although the results could be easily generalizable to the case of any positive number of firms). I devote the first section to introduce the conceptual framework of mathematical learning theory (MLT). The second is devoted to present a curious problem (that I label 'the Decodifier Paradox') that arises whenever one tries to combine the usual demand functions with the framework of MLT in order to understand the behavior of the firms in setting prices and production quotas for a particular type of good. In the third section the Decodifier Paradox is resolved, within the framework of statistical decision theory, and a solution to the conceptual problems is thereby suggested. This poses in rather precise terms the mathematical problems that have to be solved in order to elaborate continuous PFD down to the details; roughly, these problems have to do with the existence of minimizers for a certain risk function.

1. THE CONCEPTUAL FRAMEWORK OF MLT

The continuous version of mathematical learning theory is due originally to Suppes (1959). It is perhaps the first historic example of a scientific theory born to life in axiomatic form. Suppes' original formulation requires one point to be reinforced and the smearing distribution to be unimodal. It also requires the set of responses to be an interval of the real line. It turns out that these requirements are too restrictive for PFD since – as we shall see – the response of the firm consists naturally of *two* things, namely a price and a production quota, and so the set of responses is more appropriately described as a rectangle in the Cartesian plane. Also, the market and the competition may reinforce more than one response, and so it does not seem natural to require for the smearing density to be unimodal. I have modified Suppes' axioms accordingly, but their content is in spirit much the same as that of the original axioms; in fact, they are just a slight generalization of them. The axioms of MLT come in three groups dealing respectively with conditioning, sampling, and responses. I will proceed to introduce them in order to discuss their meaning in connection with the intended interpretation.

Conditioning Axioms

(C1) For each stimulus s there is on every trial a unique smearing distribution $K_s(x; z_1, \dots, z_m)$ on the rectangle $X = [\alpha, \beta] \times [\gamma, \delta]$ of possible responses such that (a) the distribution $K_s(x; z_1, \dots, z_m)$ is determined by its modes $z_1, \dots, z_m$ and its variance; (b) the variance is constant over trials for a fixed stimulating situation; (c) the distribution $K_s(x; z_1, \dots, z_m)$ is continuous and piecewise differentiable in the $m+1$ variables.

(C2) If a stimulus is sampled on a trial, the modes of the smearing distribution become, with probability θ, the points of response (if any) which are reinforced on that trial; with probability $1 - \theta$ the modes remain unchanged.

(C3) If no reinforcement occurs on a trial, there is no change in the smearing distributions of sampled stimuli.

(C4) Stimuli that are not sampled on a given trial do not change their smearing distributions on that trial.

(C5) The probability θ that the modes of the smearing distribution of a sampled stimulus will become the points of the reinforced responses is independent of the trial number and the preceding pattern of events.

Sampling Axioms

(S1) Exactly one stimulus is sampled on each trial.

(S2) Given the set of stimuli available for sampling on a given trial, the probability of sampling a given element is independent of the trial number and the preceding pattern of events.

Response Axioms

(R1) If the sampled stimulus s and the vector $z = (z_1, \ldots, z_m)$ of its smearing distribution's modes are given, then the probability of a response in the rectangle $[\alpha_1, \beta_1] \times [\gamma_1, \delta_1] \subseteq X$ is

$$K_s(\beta_1, \delta_1; z) - K_s(\alpha_1, \delta_1; z) + K(\alpha_1, \gamma_1; z) - K(\beta_1, \gamma_1; z).$$

(R2) This probability of response is independent of the trial number and the preceding patterns of events.

Since it is advisable in a first approach to the problem of applying MLT to PFD to consider the simplest cases, we shall restrict our consideration to one stimulus, which may be identified with the minimization of a certain risk function (this will be made more specific later). With this restriction, the following notation is introduced for random variables on trial n:

(i) The *response* random variable $\mathbf{X}_n$, with values x_n or simply x, distribution R_n, and density r_n;

(ii) the *reinforcement* random vector $\mathbf{Y}_n = (\mathbf{Y}_{n,1}, \ldots, \mathbf{Y}_{n,m})$, with values $y_n = (y_{n,1}, \ldots, y_{n,m})$ or $y = (y_1, \ldots, y_m)$.

(iii) the *smearing-parameter* random vector $\mathbf{Z}_n = (\mathbf{Z}_{n,1}, \ldots, \mathbf{Z}_{n,m})$ of the stimulus s, with values $z_n = (z_{n,1}, \ldots, z_{n,m})$ or $z = (z_1, \ldots, z_m)$.

(iv) the *effectiveness-of-conditioning* random variable $\mathbf{F}_n$, with value 1 for effectiveness and 0 for non-effectiveness; the probability of value 1 is θ (*see Axiom C*2).

The random variable $\mathbf{X}_n$ has a continuous distribution. In usual applications of the theory $\mathbf{Y}_n$ and $\mathbf{Z}_n$ are also supposed to have continuous distributions, but this does not seem plausible in the present situation for reasons that will be apparent later. The variable

$\mathbf{X}_n$ takes its values on X, whereas both $\mathbf{Y}_n$ and $\mathbf{Z}_n$ take theirs on X^m. The fourth variable, $\mathbf{F}_n$, has always a discrete distribution.

In a typical learning sequence, the subject (of the experiment) is presented on trial n with the stimulus s and responds choosing a point x_n in the rectangle X of responses, according to the probability distribution $K(x; z_1, \ldots, z_m)$ with density $k(x; z_1, \ldots, z_m)$ and modes $z_1, \ldots, z_m$. Then the reinforcements $y_{n,1}, \ldots, y_{n,m}$ occur and therewith two possibilities arise: Either (1) $y_{n,i}$ (for some $i = 1, \ldots, k$) is close or equal to x_n in which case, by Axiom (C2), the smearing distribution K remains conditioned to s; or (2) the reinforced responses $y_{n,i}$ are quite different from x_n. If the reinforced responses $y_{n,i}$ are all distinct from the response x_n, then with probability $1 - \theta$ the stimulus remains conditioned to K and, with probability θ, the $y_{n,i}$ become the modes of a new smearing distribution $K'(x; y_{n,1}, \ldots, y_{n,m})$ that has the same variance as K but different modes. θ is a number between 0 and 1 called the *learning parameter*.

The aim of PFD is to construct the sequence of quarterly decisions of a firm regarding the choice of a price and a production quota as a learning sequence in the above described sense. Roughly, supposing that there are two firms, namely Firm 1 (F1) and Firm 2 (F2), any of the firms, say F1, choses on trial (quarter) n a point $x^* = (p^*, q^*)$ in a Borel subset of X with probability determined by the smearing distribution $K_n^1(x; z_{n,1}, \ldots, z_{n,m})$. As a result of this action, the market reacts demanding an amount $\hat{q}$ of goods, and then the problem consists in interpreting the reinforcements that such a demand would be determining. The conceptual problems of the theory are clustered around this question. Unlike the typical experiments, where the reinforced response is an easily observable event, like the occurrence of a point of light on the rim of a disk, in the present application the reinforcement or reinforcements are not as easily observable, and some theoretical apparatus is required in order to decodify the observed events and find out the reinforcements they are actually pointing to. Naive attempts to solve this riddle give rise to a very strange situation that I have labeled the Decodifier Paradox. To this I turn now.

2. THE DECODIFIER PARADOX

As I said before, the response of Firm F1 on trial n consists of setting a price p^* and a production quota q^*, i.e., of choosing a point

$x^* = (p^*, q^*) \in X$, according to a probability distribution determined by a density $k_n^I(x; z_{n,1}, \ldots, z_{n,m})$, where the $z_{n,i}$ are the unique points where k reaches a maximum on X. The actual reaction of the market is an observable demand $\hat{q} \in Q = [\gamma, \delta]$, but this immediately poses a problem for the theory, since the set of reinforcements R has to be equal to the set of responses X, i.e., $R = X$. The only possible solution to this problem seems to consist in the assumption that the actual demand is in fact pointing out to the "correct" pair or pairs $(p, q) \in X$, but this involves the idea that the actual demand is something like a codified message. The problem that arises with this idea is that in order for the firm to receive the reinforcement that the actual demand would be determining, the firm needs a decodifier to interpret the message. In what could this decodifier consist of? The resource that presents itself as a natural candidate is a function correlating demands with prices, i.e., a so-called Marshall function, which is a function assigning a price to each demand. Unfortunately, such a resource cannot be used in this case, because Marshall functions represent good correlations between demands and prices (if at all) only under monopolistic conditions, and PFD intends to be applied to competitive markets in the first place. Moreover, even if we were able to find a way to overcome these limitations of the Marshall functions, the problem would stand anyway, for the following reasons.

Suppose that the firm believes that the demand function[1]

$$q = \varphi(p) = e^{\sigma} p^{\eta},$$

where η is the elasticity of the demand, is a good estimate of the correlation between prices and demand. It then follows that the following functional relation holds as well:

$$p = \varphi^{-1}(q) = e^{-\sigma/\eta} q^{1/\eta}.$$

Thus, φ^{-1} would be the required decodifier, since it can be used to infer that the reinforced point at demand $\hat{q}$ is $(\varphi^{-1}(\hat{q}), \hat{q})$. Is this really so? Actually, in using φ^{-1} as a decodifier the firm is assuming that φ^{-1} is accurate enough. But this is tantamount to assuming that the inverse φ is equally accurate, and so it seems that in order for the firm to have a means of interpreting the behavior of the market, it must also have a means of estimating the possible demand given a certain price.

A curious paradox – which I want to label 'the Decodifier Paradox' – is generated by the situation just described. Suppose that using φ the firm fixes a price p and a production quota q which is close to $\varphi(p)$, the predicted value of the demand according to φ. What happens if the actual demand is far removed from $q = \varphi(p)$. Isn't it an actual dismissal of φ as an accurate demand function? And, if this is so, how can the firm rely at all on φ, or its inverse φ^{-1}, to interpret the meaning of the actual demand for future price determination? It would seem that the very lack of coincidence between actual and expected demand leaves the firm with no way of figuring out what correction in these previsions is the actual demand suggesting.

In view of the former difficulties, it is advisable to attempt another approach to the situation faced by a firm (Firm 1) at the beginning of the quarter, under conditions of competition with another firm (Firm 2) producing the same kind of good. Typically, Firm 1 (the same can be said, symmetrically, about Firm 2) may have some estimate of the demand for the good within the population target of the same, or may attempt to obtain a certain response by means of an advertising campaign that takes into account the relevant cultural and economic factors. Also, Firm 1 must take into account the possible behavior of Firm 2. The name of this competitive game is to sell as many goods as possible at the highest possible price; the most desirable situation for Firm 1 is to push Firm 2 out of the market altogether, and to have the highest possible demand at the highest possible price. Assuming that there is no collusion between the firms, i.e., that the market is a real duopoly, Firm 1 decides to set price p^* for its good, hoping that the similar decision of Firm 2, as well as the conditions of the potential buyers, will allow Firm 1 to sell an amount q^* of goods. Thus, Firm 1 decides to set price p^* and to produce exactly q^* goods. It appears clear at this point that in a natural, intuitive sense the response of Firm 1 consists precisely of choosing *both* the price p^* and the production quota q^*, i.e., the point (p^*, q^*) in the space X of all such possible pairs.

Suppose that one quarter later Firm 1 decides to evaluate its first decision. Basically, there are three possible outcomes: The actual market demand $\hat{q}$ was *less* than q^*, *equal* to q^*, or *greater* than q^*. It is clear that if $\hat{q}$ is less or greater than q^*, then the firm *should* make some adjustments, i.e., such outcome counts as negative reinforcement. The problem is, again, that it is far from clear how to interpret this negative reinforcement. One possibility is to take that outcome as reinforcing a modification of the original product quota to meet the actual demand, i.e., as reinforcing the point $(p^*, \hat{q})$ where p^* is the same price as before, but if this were done systematically the price p^* never would get changed, only the estimate of the associated demand. This situation is not only fatalistic but also unrealistic. In fact, the actual demand's being distinct from the production quota can also be interpreted as reinforcing a modification in the original price. It is plausible to assume that the reinforced price is greater (less) than the original one if the actual demand is greater (less) than the chosen production quota, but the theory requires much more than just this reasonable remark: The theory requires a finite number of points to be reinforced, whereas the remark leaves open the possibility of there being a *continuum* of points of reinforcement. Which, among all these points, are the ones that are being reinforced? How far should this point be from the original price p^*? The answer seems to be: That depends on the magnitude of the difference $|\hat{q} - q^*|$, i.e., the greater such difference is, the farther must be the reinforced price from the original one. But this is not necessarily so, because the actual demand for Firm 1 does not depend exclusively on the price set by the same firm; it depends also on the behavior of the other firm. For instance, consider the following situation. Imagine that on quarter n Firm 1 sets price 5 for its goods and expects that his share of the market at this price will be 100 items, guessing that the profit rate of its competitor, Firm 2, as well the latter's production costs, are similar to its own. Nevertheless, Firm 2 decides to push its competitor a little bit by setting a price of 3, slightly above the production costs, in order (say) to implement an advertising campaign. As a result of this move, Firm 1's actual demand is reduced to only 50 items, since Firm 2 gets a larger share of the market. Which should be the behavior of Firm 1 for quarter $n+1$? Again, which is the reinforced price, the reinforced demand, or both? It seems clear that a different approach is needed, an approach taking into account not only estimates of the disposition to buy that consumers will have at quarter $n+1$, but also the behavior of Firm 2 in the same quarter.

3. A DECISION-THEORETIC APPROACH

The approach that I shall propose here is based upon the idea of a probabilistic demand function for a firm. A probabilistic demand function for Firm 1 is a conditional density $d^1(q|p_1, p_2)$ which gives the probability of demand q for F1 belonging to a Borel subset of Q, given that F1 chooses price p_1 and F2 chooses price p_2. If continuous PFD intends to be an empirical theory, it would have to adopt as an empirical claim that Firm 1 actually possesses a probabilistic demand function of the sort described and, additionally, that it also possesses an estimate of Firm 2's propensity to choose a price i.e., a density $e^1(p_2)$ over $[\alpha, \beta]$. Under this assumption, F1's decision problem consists in deciding, out of the information afforded by d^1 and e^1, which price and production quota to choose for the current quarter.

This problem is better posed in terms of a game "against nature". Let us consider the possible demands $q \in Q$ as possible "states of nature" and the elements x of X as "actions". The "statistician" (Firm 1) chooses an action x in X without being informed of the point that "nature" (the market) is going to choose; moreover, the market choice depends in part on the choice made by Firm 1, but this firm (as well as the other) has only an estimate of the demands that may ensue if it chooses a given price. It is necessary also to assume that Firm 1 has a "loss function", which is a real-valued function $L(q, x)$ defined on the Cartesian product $Q \times X$. In this way, a game $\langle Q, X, L\rangle$ is defined, although the loss function L has not been defined yet.

L may be defined in many ways, and I do not think that PFD should be committed to a particular way of defining it, although a typical way of characterizing it is as a relation between the actual revenue and the expected one. Thus, if (p^*, q^*) is the action chosen by F1, and q is the actual demand, the loss function may assume the following form:

$$L(q, (p^*, q^*)) = \lambda(p^*q - p^*q^*)^{2\nu}$$

where λ is positive and ν is a positive integer. It must be left clear, nonetheless, that PFD does not require L having this specific form, and it is preferable to make room for different applications of the theory having different loss functions.

Making use of its estimates $d^1(q|p_1, p_2)$ and $e^1(p_2)$, Firm 1 may determine its risk functions as follows. First of all, it can obtain a conditional density giving the probability of a certain demand occurring, given that Firm 1 chose price p_1, since this is none other than the integral

$$d^1(q|p_1) = \int_\alpha^\beta d^1(q|p_1, p_2)e^1(p_2)\, dp_2. \tag{1}$$

If F1 had a probability density determining its own propensities to choose a price, then it would be able to obtain an unconditional probability on the possible demands by means of the right hand side of Equation 1. But it is not natural to expect such a thing because in fact F1 has to *decide* which price to choose and this action is under its complete control; in other words, F1 does not need a probability distribution on its possible choice of prices, but rather a means of determining what is the best price choice according to some criterion. This criterion is provided by the risk function, which is obtained as follows. Notice that if F1 choses action $x^* = (p^*, q^*) \in X$, then p^* certainly occurs and so, under the assumption that this action has taken place, the probability density over the demands for F1 is just $d^1(q|p^*)$. Hence, the value of the risk function ρ at point x^* is given by

$$(2) \qquad \rho(x^*) = \int_{\gamma}^{\delta} L(q, (p^*, q^*)) d^1(q|p^*)\, dq.$$

Since the minimizers of the risk function ρ are naturally interpreted as the reinforced responses for F1, MLT requires the number of these minimizers to be finite and equal to a constant number on each trial. The theory would be more manageable if it could be shown that there is precisely *one* minimizer on each trial, but this is not absolutely required. It turns out that the main mathematical problem that continuous PFD gives rise to is precisely this, namely, *to determine which conditions must be imposed on the densities d^1 and e^1 in order to guarantee the existence of a positive integer m such that, for every pair of such densities, the risk function ρ obtainable from them in the prescribed form has exactly m minimizers.* If this problem were solved, the minimizers $z_1, \ldots, z_m$ of ρ could be identified with the modes of the smearing distribution K on the current trial.

For the sake of theory construction, suppose now that all estimates d^1 and e^1 determine risk functions ρ having precisely m minimizers, and let d_n^1, e_n^1 be the estimates of firm F1 on trial n. If $z_{n,1}, \ldots, z_{n,m}$ are such minimizers, with probability given by $K_n^1(x; z_{n,1}, \ldots, z_{n,m})$ F1 chooses action $(p^*, q^*) \in X$. If the actual demand $\hat{q}$ on that quarter for F1 is equal to chosen production quota q^* then, by the Error Learning Rule,[2] according to which the estimates d^1 and e^1 should not be changed on trial n if the forecast they predicted turns out to be correct, F1 maintains its basic estimates, which of course lead again to the same points of reinforcement. Thus, in this case positive reinforcement occurs and by Axioms (C1) and (C2) K_n^1 remains as the smearing distribution with probability 1. If, on the other hand, the actual demand $\hat{q}$ is removed from q^*, then negative reinforcement occurs and a new pair of estimates is suggested, in which case two things may happen: Either (1) F1 sticks to its original estimates d_n^1 and e_n^1 with probability $1-\theta$, and so with this same probability the modes of the smearing distribution, i.e., the minimizers of the risk function determined by d_n^1 and e_n^1, remain unchanged; or (2) with probability θ F1 takes the market response as requiring the adoption of new estimates (one of them or both), which lead to a new risk function having different minimizers $y_{n,1}, \ldots, y_{n,m}$, in which case with probability θ these points $y_{n,1}, \ldots, y_{n,m}$ become the modes of a new smearing distribution $K_{n+1}^1(x; z_{n+1,1}, \ldots, z_{n+1,m})$ (where $y_{n,i} = z_{n+1,i}$ for $i = 1, \ldots, m$).

In order to provide an interpretation of all the primitive parameters of the theory, we still have to consider the meaning of $\mathbf{Y}_n$, the response random vector, and that of $\mathbf{Z}_n$, the smearing-parameter random vector. In controlled experiments, both variables have continuous distributions and $\mathbf{Y}_n$, that has distribution F_n and density f_n, is controlled by the experimenter. In other words, in such cases the experimenter determines the probability of the reinforcement falling into a particular Borel subset of X^m. The situation is different when a real market is being considered. In this other case nobody knows in advance which points are going to be reinforced on any trial without previously knowing which estimates is the actual demand suggesting but, once these estimates have been determined, the probability of the corresponding risk function minimizers' being the reinforced responses is 1. What this means is that f_n is in fact a degenerate joint probability distribution assigning probability 1 to the vector of minimizers of the risk function determined by the estimates suggested by the actual demand on trial n, after the reinforcement has taken place, i.e., if $y_n = (y_{n,1}, \ldots, y_{n,m})$ is the set of such minimizers, then

$$f_n(y_n) = \Pr(\mathbf{Y}_{n,1} = y_{n,1}, \ldots, \mathbf{Y}_{n,m} = y_{n,m}) = 1.$$

The random vector $\mathbf{Z}_n$, on the other hand, gives rise to a joint probability distribution g_n determined by the following theorem:[3]

(3) $$g_{n+1}^1(z) = (1-\theta)g_n^1(z) + \theta f_n^1(z)$$

and the original distribution g_1^1. In terms of formula (3), we can think of densities g_n and f_n as follows. On the first trial, F1 adopts a pair of estimates d_1^1 and e_1^1, whose corresponding risk function has minimizers $z_{1,1}, \ldots, z_{1,m}$. Thus, one this first trial the probability of such numbers being the modes of the smearing distribution K_1 is 1. On trial n, the distribution g_n is determined by means of the recursive relation (3). Notice that in any case g_n is a joint probability distribution which assigns nonzero probabilities only to a finite number of vectors in X^m.

The remaining random variables, the response random variable, whose density is r, can be obtained by means of the General Response Theorem:[4]

$$r_n(x) = \sum_{i \in I} k_n(x; z_i) g_n(z_i),$$

where $\{z_i\}_{i \in I}$ is the set of all vectors in X^m such that $g_n(z_i) \neq 0$; as I said before, this set is always finite. In this form, all the parameters of the theory have received an interpretation, to configurate a story that seems to have some economic-theoretic sense. The problems that are left open are basically two: (1) To investigate which conditions on the estimates d^1 and e^1 guarantee the existence of a finite specified number of minimizers for the corresponding risk function; and (2) to give a precise meaning to the idea of an actual market demand "suggesting" a particular pair of estimates. It seems to me that only after these problems have been solved the problem of finding out the conditions leading to market equilibrium can be adequately addressed.

NOTES

* I am glad to acknowledge the support given to the research project leading to the present paper by the National University of Mexico, through the DGAPA, as well as by Tilburg's Catholic University (Holland). I appreciate also the fruitful comments and suggestions made to me by Dr Gustavo Valencia (*Facultad de Ciencias*, University of Mexico) in connection with some of the topics presented in this paper.

1 According to Lange (1962), in actual practice demand functions almost always must be taken as having the form adopted here.

2 Fuchs (1979) formulates the Error Learning Rule as follows: "The expectation function ψ_{t-1}^i of an agent i at some time $t-1$ should not be changed at time t if the forecast it has predicted turns out to be correct". Fuchs' expectation function is different from our estimates, but there is an obvious analogy. In the paper referred to here, Fuchs investigates whether error learning behaviour is stabilizing and concludes

that in general this is not the case. Since his conceptual apparatus is so different from the one used here, his results are not easily translateable to results in PFD. Indeed, much research would be needed in order to use his results in this field since, for one thing, his results do not even employ probabilistic concepts.

[3] See Suppes (1959), p. 354.

[4] Suppes (1959), p. 352.

REFERENCES

Fuchs, G.: 1979, 'Is Error Learning Behaviour Stabilizing?', *Journal of Economic Theory* **20**,

Lange, O.: 1962, *Introduction to Econometrics*, Pergamon Press, Oxford.

Suppes, P.: 1959, 'Stimulus Sampling Theory for a Continuum of Responses', in K. J. Arrow, S. Karlin, and P. Suppes (eds.), *Mathematical Methods in the Social Sciences*, Stanford University Press, Stanford.

Suppes, P. and Atkinson, R. C.: 1960, *Markov Learning Models for Multiperson Interactions*, Stanford University Press, Stanford.

Suppes, P. and Carlsmith, J. M.: 1962, 'Experimental Analysis of a Duopoly Situation from the Standpoint of Mathematical Learning Theory', *International Economic Review* **3**.

Manuscript received 25 January 1988

Instituto de Investigaciones Filosóficas
Universidad Nacional Autónoma de México
México, DF 04510
México

MARIA ROSARIA DI NUCCI PEARCE AND DAVID PEARCE

ECONOMICS AND TECHNOLOGICAL CHANGE: SOME CONCEPTUAL AND METHODOLOGICAL ISSUES

OVERVIEW

Economic analysis has given rise to several conflicting accounts of technology and of the rate and directions of technological change. In this paper we examine some of the contrasting images of technology that have arisen in economics and we discuss some of the conceptual and methodological questions connected with the study of technological change (TC for short). We argue for a microeconomic approach in which TC is considered against the background of industrial, institutional and market structures. But we suggest that attempts to introduce into this framework cognitive models of *scientific* progress are doomed to failure, because of the fundamental differences between scientific and technological knowledge and the basic disanalogies between TC and scientific progress. In particular, we argue that the efforts of Dosi (1982), (1984) and others to treat technology and TC in a Kuhnian framework, by applying notions like ***technological paradigm***, ***normal technology***, and ***technological revolution***, are misleading. By contrast, we hold that, given the influence of economic markets, industrial and institutional structures on the development of technology, it is more plausible to regard TC as a continuous and incremental process, rather than as suffering Kuhnian crises and revolutions.

The paper is organised as follows. Section 1 introduces some basic concepts needed for the analysis of technological change. Section 2 contains some general remarks on technology in economics and reviews some of the main macroeconomic growth theories. In Section 3, we turn towards the more 'applied' perspective of microeconomic analysis, within the theory of the firm and industrial organisation, with emphasis on the recent work of Nelson and Winter (e.g., 1982). In Section 4 we deal with the conceptual relations between science and technology and the general question of the applicability of Kuhn's model of scientific change to TC. This serves as a basis for examining in Section 5 Dosi's approach to TC which embeds elements of

Erkenntnis **30** (1989) 101–127.

microeconomic analysis within a Kuhnian conceptual framework. Lastly, in Section 6, we discuss what seems to be a basic principle governing TC which we call the principle of industrial-technological continuity.

1. TECHNOLOGY: SOME BASIC NOTIONS AND PROBLEMS

The study of technology and technological change is a hybrid discipline that provides a natural meeting ground for philosophers, economists, sociologists and historians of the engineering and applied sciences. This is all to the good. Interdisciplinary research and the pooling of methods and expertise from different fields should lead to a better understanding of technological progress and its impact on society. Yet because of its hybrid nature, the study of technology is characterised by distinct and often contrasting research traditions. They diverge not only in their methodologies, but in their basic terminologies too. Even within a single discipline like economics, conceptual differences are striking, beginning with the notion of technology itself for which there is no universally accepted definition. Sometimes even compatible accounts of TC look very different from one another because they employ the concept of technology in a wider or narrower sense.

Starting from the original meaning of technology as a body of knowledge about techniques, we can regard technological change as consisting of new knowledge about such techniques, and think of technological progress as comprising a special case of technological change. Following Freeman (1979), we distinguish *technological* from merely *technical* change, since the latter need not involve essentially new knowledge, but may refer simply to the adoption or diffusion of existing or improved techniques.[1]

Borrowing a now standard classification first used by Schumpeter, TC can be analysed through the sequence *invention – innovation – diffusion*. The invention phase can be seen as related to the sphere of R&D. Research is directed at the enlargement of present knowledge and can be subdivided into *basic* and *applied* research. The stage of *development* deals with the application of research results (e.g., for the construction of prototypes and models), and should lead to an extension of the technical horizon, or of the technology, considered as the state of technical knowledge of an economy. The phase of innovation then represents the first economic application of inventions, or of new techniques, and one can distinguish here between *product* and *process* innovations. Diffusion can be considered the final stage in the process of technological change, and refers to the spreading of innovations sectorally, internationally, and so on.

To obtain a clear grasp of the determinants of technological change, all these aspects and phases must be taken into account, preferably as a unified whole. Thus, research in basic and applied science may lead to *scientific* progress, the development stage may represent *technical* progress; and whilst the two together do not imply TC, they do determine what might be called a *potential* technological change. However, since the real core of TC involves the introduction and diffusion of innovations, in general the matter of 'progress' is extremely complex. In economic theory, for instance, the analysis of technology and technical progress has been and still is a matter of dispute and controversy. In both macro and microeconomic analysis, the basic categories, hypotheses and explanations offered are loaded with numerous, often debatable, assumptions and are subject to conceptual, theoretical, and methodological problems. Even today there is no single integrated corpus that might be termed an economic theory of technological change.

Lacking a well-defined economic theory of TC, many of the disagreements among economists arise already at the foundational level of the discipline. Consequently, they involve issues of the kind that are of central concern also to philosophers and historians of technology. Important here are questions surrounding the conceptual relations between science and technology: how, for instance, technological knowledge is to be distinguished from scientific knowledge, and how the aims and patterns of technological progress are related to those of scientific progress. Both economists and historians of technology have increasingly begun to borrow instruments and models developed within the history and philosophy of science. For example, in the two-and-a-half decades since the first edition of *The Structure of Scientific Revolutions*, Thomas Kuhn's concepts of scientific community, paradigm and revolution have been transferred to an ever-wider range of disciplines; and historians and philosophers of technology have been drawing on Kuhnian concepts in analysing the problems of technological knowledge and its growth. Nowadays also economists, when studying the phenomena of innovation and technical change, have turned to the use of notions like technological community, technological paradigm and technological revolution. And, just as in recent years science studies have focussed increasingly on the contrasting images of scientific change as cumulative and continuous as opposed to noncumulative and discontinuous, so a similar development can be observed within technology studies.[2]

2. TECHNOLOGY AND TECHNICAL CHANGE IN ECONOMIC THEORY

Technological progress has long represented a shadow zone of economic theory in its efforts to explain economic growth. Apart from isolated instances, economic analysis before Schumpeter showed little interest in technical change. Although the classical economists did not ignore it, and Adam Smith's introduction of the concept of labour division was to be a milestone, under their construal economic growth is determined by labour, capital and land and less by technical progress, considered as a variable of extra-economic origins. In Marx one finds a first systematic analysis of the effects of technological progress on employment and on the development of the capitalist system, but the impact of this account of the evolution of other economic schools was to be limited, and the dominant neoclassical analysis was to offer a static interpretation of the process of factor combination, where technology is given as a datum. Technical change was to be represented by a shift of the isoquant or by changes in the transformation curve, and no explanation for the rate and direction of TC was supplied.

In Schumpeter's synthesis of economic growth and fluctuations, one recognises for the first time technological innovations as the motor driving economic growth: without innovations an economy would reach a static equilibrium position of a circular flow. In Schumpeter's view, technological change is assumed not to be steady and even, but to come in waves. There is no single factor of causation, since a great diversity of events may affect the form and timing of the waves. Innovations are uneven at any one time, exhibiting a tendency to concentrate in certain sectors; and they are not evenly distributed over time, i.e., they have a tendency to cluster and bunch simply because first some, then most, firms follow in the wake of a successful innovation. Remarkable here is the departure from the neoclassical postulate that perfect competition is beneficial for the innovative climate and monopoly is detrimental. Set against this is a new view of competition based on innovation, which stresses the creative role of the entrepreneur. Temporary monopolistic profit (real or expected) deriving from pioneering a new process/product is the major element pushing firms to innovate (the so called 'Schumpeterian Hypothesis').

Schumpeter's account is not presented as a formal model, and many economists have lamented the descriptive rather than analytical nature of its theory.[3] Nonetheless, his contribution to the understanding and explanation of TC and the role of innovations as disruptive of existing equilibria was to offer thirty years later a fresh impulse to much of the literature (more on the empirical than on the theoretical side) known today under the label of neo-Schumpeterian.[4]

Because of their implications for growth, from the early 1950s technology and technological progress began to attract a wider interest and to be more fully covered in macroeconomic analysis, where technology is treated within the framework of growth theory in terms of its implications for economic growth. The first empirical study explaining the importance of technical progress for economic growth was based on the measurement of global productivity. Solow (1957) estimated the efficiency parameter of a Cobb–Douglas production function for each year in the period 1909–49, and calculated the growth rate of this parameter as a residual which he called 'the growth rate of the technology'. After this pioneering attempt, various different studies, in the framework of what could be designated neoclassical growth theories, have followed.[5] There, in general, given an aggregate production function, of the type $Y = F(K, L, t)$ ('K' standing for capital, 'L' for labour and 't' for time) where output is linked to inputs and time, technological progress is represented by a *shift in* or *along* the production function, and by increases in the marginal and average productivity of at least one factor of production.

The main methods of classifying the causes and effects of technological progress (TP) focus on the neutrality or bias of TP, and on its endogenous or exogenous character. TP has been defined as *neutral* when it does not 'disturb' the variables of the system (in accordance with the neoclassical position, in Hicks' model, for example, when capital intensity and the marginal rate of substitution do not change), or as *non-neutral* when it has a specific factor bias (labour saving in

Hicks' model when, with a constant labour intensity, the marginal productivity of capital grows more than that of labour).[6]

Technological progress has been considered ***endogenous*** when it is caused by new capital or labour. So the efficiency of the production process increases in time dependently of, for example, capital accumulation. It is *exogenous* when the introduction of new and more efficient production processes is determined from variables external to the model, falling like manna from heaven. Thus 'time' in Solow's model plays a key role, almost as a factor of production. In this case the attempt to estimate the residual component of aggregate production functions forced a departure from static production functions (Cobb–Douglas) in favour of dynamic functions (the Solow type $Y = f(K_t, L_t, t)$), parameterised and specified through four parameters: efficiency, distribution, substitution and homogeneity.[7]

Other attempts, also within the neoclassical school, to enlarge the range of aggregate PFs by introducing the case of variable coefficients, or the decomposition of the residual factor, have still not been fully convincing. If these approaches have been subject to stark criticism on theoretical grounds, it is however the empirical content of the neoclassical production theory, especially the estimate of the aggregate production function, that has been more markedly prone to attack. In spite of their individual differences, the growth models, whilst on the one hand avoiding some of the constraints typical of static neoclassical analyses, are, on the other hand, still heavily burdened by the use of an adapted neoclassical tool-box and by the aggregation problem. Furthermore, they remain preoccupied with explanations based on equilibrium analyses.

The post-Keynesian school, by stressing the artificial nature of the distinction between movements of and along the production function, has counterproposed to the neoclassical PF a technical progress function (Kaldor) conceived as an endogeneous variable, in which the labour productivity changes depend on changes in the investment rates. Here, however, the technical progress function does not distinguish between technical and economic factors.[8] Productivity changes are effected by increases in the capital intensity. The neutrality of technical progress, unlike Hicks's neutrality, is determined by the fact that it calls for a rate of capital change equal to the rate of change of product generated by it. According to this approach, therefore, technical progress is exclusively determined by capital.

The construal of TP as endogeneous has also given rise to various different models. Their only common ground is that new knowledge is taken to be an outcome of the production process, so that the accumulation of knowledge depends on the economic variables of the system. As such they lay greater emphasis on price factor induced (Hicks), income factor induced, investment induced (Kaldor, Arrow), research induced, and demand induced technological progress.

Other models maintain a clear distinction between what could be called the realisation and the cause (endogenous vs exogenous) of TP. Approaches differ according to whether technical progress is taken to be *disembodied* from the factor inputs (i.e., it is generated when the external accumulation of knowledge is directly transferred to the production process) or *embodied*. In the former case three basic standard forms of technological progress emerge. In the first, the rate of TP does not affect the marginal rate of factor substitution (corresponding to the Hicks' neutral TC). The second suggests a TC that makes capital more efficient even if change is not embodied in new types of capital (the Solow-neutral TC). The third considers TP as increasing the effectiveness of labour in a similar manner to Solow's capital augmenting TC (Harrod-neutral TC). The embodiment approach, where technical progress is a qualitative component of the production factors, has given rise to two different models: the vintage models and the human capital models.

On the whole, growth theory deals more with the effects of technological progress on macroeconomic variables, than with its causes. Progress remains relegated to process innovation, and the determinants, rate and direction of TC, as well as the influence of the technical and scientific environment, are questions left very much in the dark. Though the growth theory perspective still counts on some supporters, in general it would seem that this bundle of analyses, albeit enlarging economists' awareness of the 'technology factor', has provided dubious results backed by little conclusive evidence. The failures of the attempts to explain technological change in the framework of equilibrium models (also within the post-Keynesian tradition) are only too apparent. It would seem that much theoretical effort has taken us little further than Solow's original dictum (1957) that technical change is "a short hand expression for any kind of shift in the production function". Whether such accounts represent a theory at all remains a matter of doubt and speculation. In this respect, it is microeconomic analysis, along with more eclectic approaches, that, by pinpointing the causal relationship between technology, the rate and direction of technical change and economic indicators of structure and performance, have offered more satisfactory answers. To quote Nelson and Winter (1974), "... research on economic growth within the neoclassical theory is creating new intellectual problems more rapidly than it is solving them. One can continue to search for solutions for these problems guided by the assumption of the neoclassical theory. Or, one can try a new track."[9] It is to these aspects that we now turn.

3. RECENT TRENDS IN MICROECONOMICS

In the last two decades, several microeconomic approaches have been developed to provide a theoretical framework for the analysis of technological innovation, attributing to technology a crucial role in affecting both the performance of individual firms and the growth of the industry. Research has focussed on the determinants and the nature of technological progress and of the dynamics of innovation in the framework of market processes. Although most of these studies are still firmly in line with neoclassical assumptions, the striking difference between them and the macro-analyses is that the former examine market processes rather than market equilibria. The innovative activity is a parameter of competition, and inventions and innovations are seen as a fundamental source and cause of growth. No homogeneous theoretical corpus has emerged, but similarities, especially with regard to the fundamental role of the interplay between technology, industrial structure and industrial growth, make it possible to talk about a technology theory of the firm and industrial growth.[10]

The innovative process has been construed in terms of a demand-supply scheme, in which the firms' demand for innovation is linked to the evolution of the demand for particular products/processes and the availability of technical knowledge. Within this account, a neo-Schumpeterian approach can be discerned, centred on the role of innovations as a factor determining temporary monopolies, and characterised by the causal link between innovation, increased rates of growth and blossoming profits and market shares.[11]

Loosely speaking, analytical studies of the determinants of innovations have until a few years ago taken two distinct lines of approach, that can be classified as ***demand pull*** and ***technology push*** respectively. According to the former it is demand, and thus investments, that exert a pull on the development of innovations. By contrast, on the second view it is technological opportunity which determines innovations. Both accounts, which differ in their construal of technology as either an exogenous or an endogenous variable, are open to criticism. Demand pull fails to explain the timing of innovations and the presence of technological divergencies and apparent discontinuities, whereas technology push underestimates the extent to which economic factors influence the rate and direction of technological change. This gap has in some respects been bridged by some intermediate approaches which support the existence of both effects.

Though Schumpeter was unclear about the endogenous or exogenous character of innovation with respect to the economic structure, for the neo-Schumpeterian approach the innovativeness of the firm (as the locus of innovation), expressed through its levels of R&D and patents, is correlated to variables related to oligopolistic profit and growth and with its success in selecting optimal strategies given a set of productive resources and against the characteristics of the market in which the firm operates. However, this theory takes technological change as a variable entirely endogenous to the firm or the sector. To achieve an adequate grasp of the role of a technology in shaping the character of a sector, one has to depart somewhat from this approach and bring the influence of those exogenous factors, which were not underplayed by Schumpeter himself, firmly to the fore. This is the tendency that emerged from critical revisions proposed in the mid-1970s, when a redefining of some elements of the theory of the firm led to a different stress on phenomena such as the impact of technical change on the strategies and growth paths of the firm. Among the most significant contributions of the period are the efforts of Nelson and Winter (1977, 1982) to develop an analysis of the plurality of factors influencing R&D decisions and to lay the foundations for an evolutionary theory of technological change. Largely as a result of their work, the upshot of the technology debate since the '70s has been to underline the existence of a dynamic relationship between the economic system and technical change, based on an evolutionary process in which technological change, firms' strategies and industrial structure mutually interact.

3.1. *Nelson and Winter's approach*

Nelson and Winter's evolutionary model analyses the pattern of firms' conduct and the rate of growth of firms in conjunction with their degree of innovativeness. Innovations are no longer considered as an act of skill, but as an act of insight, and the choice of a strategy is also linked to the size of the firm and its share of the market. Their model provides a picture of an evolving industrial structure under condition of technological change. However, it would be better to talk about N&W's 'approach', rather than 'model', since in the last ten years they have offered several related but slightly different models to explain the rate and direction of TC in a context where market structure is an endogenous variable. Broadly speaking, the sets of assumptions, structural variables and behavioural rules concerning the firms' R&D strategy are roughly similar in these models, but interesting variations do occur. In one model (1982), for instance, a 'cumulative technology' is considered whereby firms innovate through incremental improvements of an available technique, without drawing on knowledge generated from outside. In other formulations, innovation is science-based and is strictly linked to the latent productivity of the firm.

On several grounds their approach can be considered a synthesis of the behavioural models of the firm and the Schumpeterian accounts. From the behavioural approach, of which Winter has been a keen advocate, they freely borrow concepts like 'changing environment' and 'routine'. The firm 'searches' its environment to obtain fragments of knowledge to improve its routines. Markedly Schumpeterian is their rejection of maximising rationality and equilibrium, their opposition to the neoclassical static efficiency of competition, and their idea of competition as a process rather than a state, where innovations play a major role in disrupting previous equilibria. But, unlike Schumpeterian approaches, their models do not incorporate elements of discountinuity and disequilibria, rather they draw on several analogies with biological evolution. Thus, the role played by mutation in evolutionary theory is comparable to the role of innovations in Nelson and Winter's models. A hint of Darwinism is conjoined with Schumpeter in their use of 'search' and market 'selection' mechanisms that generate winners and losers. Firms which have better rules for searching than others, or which introduce an innovation first, will grow more rapidly. The crucial factors leading to oligopolistic power and dynamic industrial adjustment over time are the differences (structural and behavioural) between firms.

Firms operate in a technological environment, a sort of system of exchanges of information and signals, marketwise, through which the development of scientific and technological applications are maximised and reproduced. The firms' behaviour is assessed against a background where competition is to be read as a technological rivalry. The models incorporate both internal and external behavioural rules and assumptions. Among the external conditions is *latent productivity*, defined as a source of relevant technical knowledge which, in conjunction with the application of new techniques to industry, determines the success of research efforts undertaken within the firm. Endogeneous variables determining the development path of the firm include its capital stock, capital markets and profits. The concept of 'selection environment' lends itself to the idea that the firm acts to adjust to the requirement of the technology (in terms of the rate and direction of innovation) in order to increase or decrease its minimal optimal size, to raise or lower entry barriers, to create or destroy further investment opportunities in new markets. And the firm contributes to influence technological change itself, by selecting its own pace and extent of R&D investments and by accelerating or retarding the process of diffusion of innovations.

Research efforts are related to the size of the firm and other structural variables, and to the possibility of applying presently unexploited technical knowledge. But technology is not a free good, and not all firms can gain access to technical knowledge at the same time and with the same procedures. The success of any strategy depends on access to the relevant technological potential, the latent productivity of the firm, and the extent of the appropriability of technology ahead of other rival firms. Independent of specific national differences and of public corrective action, there are cumulative effects that follow a first and fast start in the introduction of innovation. However, even firstcomers are under pressure to innovate continually, not to lose their advantage to rapid imitators. In short, technological competition between firms is considered as a sort of open market where innovation can be introduced either on the basis of internal R&D efforts or through imitation. Some firms do both imitate and innovate. However, at any potential level of technology, R&D efforts are characterised by decreasing returns to scale, since the chance to improve productive techniques decreases with the rise in achieved levels of productivity. The technological development and innovation process is then characterised by an uncertainty correlated to its degree of novelty and its technical and market results. Especially since firms do not know whether they have made the right *search* and *selection*, they may prefer to stick to a routine and incremental, as opposed to a radically innovative, behaviour.[12]

Summarising, technical progress is represented by a continuous, cumulative process of small improvements to existing techniques, as 'natural trajectories' set by the supply structure and past techniques. Every improvement represents an accumulation of theoretical and practical knowledge. Technological progress constitutes an evolutionary system. Major innovations are made possible by a set of minor, incremental changes. However, by exploiting *natural trajectories*, some firms or even whole sectors can achieve notable growth and eventually effect changes in the existing market structure.

4. EVOLUTION VS REVOLUTION. IS A KUHNIAN PERSPECTIVE APPLICABLE TO TECHNOLOGY?

Such accounts of the general characteristics of technological change readily invite comparisons with basic science and its development patterns. We are led rather naturally to ask whether the well-articulated models of scientific progress proposed by philosophers and historians of science, such as Kuhn, Lakatos and Laudan, could also be applied in analysing technology: Could such models help to explain the rates and directions of technical change? And, is technological progress well-described by Kuhnian metaphors like 'crises' and 'revolution'? In this and the remaining sections we shall try to defend a negative response to these questions, first, by examining on a general level the analogies between scientific and technological knowledge, secondly, by considering a specific, 'Kuhnian' model of TC that has been offered by Dosi, and lastly by appealing to some further traits of innovation and TC that lend weight to the 'nonrevolutionary', 'incremental' view of technological growth.

The first point to emphasise is that, with the exception of radical, sociological theories of knowledge, current theories of scientific progress are based on *cognitive* models of science which appeal to the internal standards of 'truth', 'accuracy', 'simplicity', etc., that a given scientific community upholds. These accounts of science do not deny that external factors – cultural, social, economic and political – affect the way in which basic scientific research is conducted, influencing, for instance, the choice of problems and the types of research methods. But in explaining scientific theory choice and appraising the relative merits of competing theories and solutions, it is the *internal* standards which ultimately count.

If one considers the situation with respect to technology, however, strong disanalogies emerge. Take, for example, the Kuhnian model of crisis and revolution. For Kuhn, a crisis in the process of normal scientific research is brought about by a build-up of anomalies, by the presence of an increasing number of problems that evade successful solution within the bounds of the associated scientific 'paradigm'. A paradigm will not be abandoned, however, until the basis of a suitable alternative paradigm has been formed. Moreover, when a shift to a new and revolutionary paradigm does occur, the shift is total, and 'victory' for the new paradigm is complete. In addition, the new paradigm will possess different concepts, laws and methodological standards, *incommensurable* (i.e., not logically and empirically comparable) with the old.

It is difficult to fit technology into the above pattern. For Kuhn, (revolutionary) scientific change is disruptive, discontinuous, non-cumulative. Technological knowledge, as embodied in the knowledge of techniques and processes, grows cumulatively. A technological option may come to be superseded, because a better solution is found that is more cost-effective, technically superior, or whatever. But to reject an existing (and at some time successful) technology is not to 'refute' any particular piece of technological knowledge. It is also hard to identify anything like a technological 'crisis' engendered by the internal failures of a technology. As we saw earlier, even in cases of notable technological transitions (like e.g., cable telegraphy to wireless), empirical research amply supports the thesis that the successful implementation of the new technology cannot be explained in terms of the failures of the existing technology in its 'push' to achieve higher standards.[13] Moreover, since 'rival' technologies are not constituted by rival claims to *knowledge* in the scientific sense, they are not mutually exclusive. Indeed, different technological 'solutions' may co-exist over long periods (e.g., coal vs. nuclear in electricity generation, diesel vs. petrol in internal combustion engine design, etc.). Lastly, in what sense could the 'concepts' and 'standards' associated with a new, 'revolutionary' technological paradigm be said to be 'incommensurable' with their predecessors? Insofar as the 'internal' technical parameters, like design-engineering specifications and performance indicators, are based on physically measurable characteristics, they are as stable in meaning as the underlying measurement theory permits.

Equally strong disanalogies emerge if we compare technology with another influential view of science, namely as a problem-solving enterprise.[14] Whilst both technology and science may usefully be regarded as problem-solving activities, in each case the nature of the problem-solving process is quite different. Science aims to solve *cognitive* problems by constructing explanatory hypotheses, laws and theories, whereas technology seeks to increase its store of *practical* knowledge, much of it embodied in human skills and machinery. The matter can be summed up by the dictum that science seeks solutions to *wh*-questions (why, what, when, . . .), whilst technology is characteristically concerned with *how*-questions.

In recent years, many analysts have noted correctly that science and technology are not activities *external* to the economic system, but are integral variables of that system whose role in explaining economic growth is fundamental. Accordingly, the economic study of technological change must take account of a variety of mutual feedbacks occurring along the science-technology-economics interface. At the same time, new developments in the study of the growth of scientific knowledge have tended to widen our conceptions of 'rationality' and 'progress' so as to incorporate the roles of the scientific 'community' and the scientific 'paradigm' or 'research tradition' as additional factors having an explanatory force. These new models of scientific change make science *seem* less exogenous, less autonomous than before. On a superficial examination, therefore, they *seem* to reinforce the analogies between scientific and technological forms of knowledge, and thus look promising material for the economist in search of a systematic perspective on innovation and technological change. Economists need a fluid and flexible conception of technology to do justice to the complexities of its impact on industrial and economic growth; and the current approaches to the dynamics of science appear to offer them a conceptual framework that is broad enough to accommodate technological practice, yet tight enough to yield explanatory fruit.

Nevertheless, appearances can be deceptive. From the fact that both science and technology may be intrinsic variables for and of economic analysis, we can deduce very little about their mutual interrelations. In the other direction, recognition that the basic concepts employed by the methodologist and historian of science need to be broadened and 'pragmatised' does not imply that the cognitive development of science is therefore to be explained in economic or sociological terms. Whether scientific theories are appraised with reference to their empirical adequacy, predictive success, degree of truthlikeness or their problem-solving effectiveness, the fact remains that the appraisal is a cognitive one, based on criteria that are so to say 'internal' to science itself. Despite the recent antipositivist movement in the philosophy of science, and the successive 'liberalisations' of empiricism that the work of Popper, Kuhn, Lakatos and Laudan has brought about, science remains an activity that can be and is studied from both the 'inside' and from the 'outside'. Insofar as the economist is concerned with the 'outer' dimension of science – with the mutual interplay of the scientific and economic 'systems' – the standard 'insider' analyses of scientific progress can be of little intrinsic interest to him. Insofar as these models underline only the basic disanalogies between science and technology, their adoption within the economic theory of technological change is liable to lead to error and confusion.

5. TECHNOLOGICAL COMPARABILITY AND CONTINUITY: DOSI'S TECHNOLOGICAL PARADIGMS

Among the recent attempts to develop theories of technological change that break with the neoclassical tradition in economics, several writers have urged the need to analyse technological change under some model of 'directed' and 'constrained' development, where certain kinds of development 'paths' of technology naturally arise. Thus, Sahal (1981), whose approach is inspired by systems theory, speaks of technological *guideposts* as a constraining and directing factor of change. Nelson and Winter, as we saw, take a more behavioural view (within the theory of the firm), and refer to natural technological *trajectories*; whilst Dosi (1982, 1984) also borrows the concept of trajectory, but embeds it within a markedly 'Kuhnian' account of technology. The latter approach seems to date to be the most sophisticated attempt to marry a model of scientific growth with a viable economic theory of technological change.

Dosi's conception of technology is a broad one that includes both theoretical and practical knowledge that may be of either an explicit or an implicit kind. It also includes physical equipment and devices, besides some 'perception' of the relevant scope and limits of the knowledge in question. By analogy with science, he regards a technological *paradigm* as a "《model》 and a 《pattern》 of solution of *selected* technological problems, based on *selected* principles derived from natural sciences and on *selected* material technologies" (Dosi, 1982). A technological *trajectory* is then defined as the development path of *normal* technology (by analogy with Kuhnian normal science). More specifically, he suggests that a trajectory be represented "by the movement of multi-dimensional trade-offs among the technological variables which the paradigm defines as relevant" (ibid). The variables in question will usually comprise both strictly technical quantities as well as broader economic factors; and a partial characterisation of *progress* within normal technology can be given in terms of improvements in the trade-offs between these technical and economic variables. It is important to note that a technological trajectory is not identified with a single 'path', but is viewed rather as a "cluster of possible technological directions whose outer boundaries are defined by the nature of the paradigm itself" (ibid).

One important feature that the technological paradigm is supposed to highlight is that technological change may be strongly *directed* towards certain problem areas, to the exclusion of others. In its focus on particular problems, the paradigm thus acts as a *constraint* on, as well as providing a *heuristic* for, technological development. Additionally, this conception allows for the rational comparison of technological advances within a given paradigm, and for the idea of progress – including as a special case cumulative progress – within normal technology. Moreover, *continuous* technical change is distinguished from *discontinuous* change, the former occurring in normal, the latter in *extraordinary* technology; and radical innovations mark the emergence of a new technological paradigm. Though Dosi does not insist on the total 'incommensurability' of different paradigms, transparadigm comparisons of technologies are seen as highly problematic:

it might prove impossible to compare *ex ante* two different technological paradigms and even *ex post* there might be overwhelming difficulties in doing it on solely technological grounds. (Dosi, 1982, p. 159)

Without going further into the details of Dosi's model, several questions concerning the general framework come to mind. First, there is the matter of the applicability of the paradigm concept of technology. The very fact that one can speak at all of, say, a "semiconductor" technology implies, of course, that a certain 'domain' of R&D and industrial production is at issue. A certain body of knowledge, techniques, skills and machinery is associated with this domain, as are groups of firms, institutions, products and markets. Any criterion for marking off one 'technology' from another inevitably implies that certain pieces of knowledge, certain types of problems and certain kinds of research strategies are associated with, and are to a greater or lesser extent specific to, *that* 'technology'. To individuate a technology means, in other words, to lay boundaries, set forth constraints, exclude alternatives.

Any systematic study of TC is bound, therefore, to assume that technologies can be grouped, categorised, classified, distinguished from one another. To say that these groupings and classifications should be subsumed under the broader rubric of 'paradigm' is to say very little unless features specific to the Kuhnian concept, can be, first, identified as characteristic traits of technologies, secondly, used to explain why technologies develop in some ways rather than others.

We have noted already several pronounced disanalogies between science and technology; and they should warn us against any hasty transference of metascientific concepts to the area of technology studies. Dosi himself is aware of the dangers here. He proceeds tentatively, stressing the idealised and approximate character of the paradigm notion in technology. He goes as far as to suggest that, for many purposes, one could substitute for 'paradigm' alternative concepts like 'research programme' (in the sense of Lakatos). But there is a risk here of mixing one's metaphors. For example, it is scarcely controversial to assume that research and development programmes are *associated* with technologies and technological advances; programmes that may well be characterised by a certain internal 'logic', and constrained by given technical and economic considerations. But in this sense the term 'programme' (as in, e.g., developing SDI, fifth generation computers, or the lean-burn petrol engine) is really to be understood in the sense of 'scheme' or 'project'. It may be more or less open-ended, more or less product-specific, but is at any rate something instigated by governments, institutions or firms, and represents a highly goal-directed activity. This seems to be, on the one hand, a quite different usage than is meant by 'research programme' in Lakatos' sense, and, on the other hand, a far cry from constituting anything like a technological 'paradigm' in the sense Dosi intends.

There are further difficulties. One of the attractive features of the notion of scientific paradigm is that it invites one to subsume under a single heading a whole variety of theories applying to quite different phenomena and problem-domains, but which are, in a well-defined sense, ***homogeneous***, i.e., there are strong 'family resemblances' that unite the different theories belonging to a paradigm. Quite the reverse feature seems to apply in the case of technology: a 'paradigm' in the latter sense appears to be relatively 'domain-specific' and homogeneous as far as its characteristic problems are concerned, yet quite heterogeneous and disjointed in other respects, e.g., with regard to techniques, materials, methods. This is in part due to technology's orientation towards practical knowledge, its reliance on basic and applied scientific research, and its need to borrow freely from any available source that may provide relevant inspiration of materials for tackling a given problem. This overlapping of fields and general lack of homogeneity makes the demarcation of technological paradigms difficult and throws into doubt the usefulness of this concept. As Dosi admits, it is often more correct to associate a certain field of technology with a *family* of paradigms, e.g., semi-conductor paradigms, nuclear paradigms, and so forth. Thus in the field of, say, electricity generation, one must first distinguish 'nuclear' from 'oil', 'coal' and others. Within nuclear one must further distinguish the LWR technology from alternatives like that of gas-cooled reactors or FBR. And within the LWR 'paradigm' further clear distinctions can be made, e.g., the separation of BWR from PWR technology. At what point in this classificatory scheme one can be said to have reached a natural 'paradigm' is far from obvious, even though all the technologies concerned are very closely tied to a particular 'product', the nuclear power plant.

The idea of paradigm-bound technology raises further questions regarding technological progress. In the first place, since progress is always a matter of trade-offs between different variables (and even quite distinct kinds of variables), TC along a single path will seldom represent 'linear' or cumulative progress, since improvements in some relevant parameters will often be bought at the cost of accepting disadvantages in others. The matter is further complicated when we consider the movement of a technological paradigm along some trajectory. Since some paths in the trajectory may advance, where others decline, the progress of a paradigm is also a matter of trade-offs, now of a 'higher-level' sort. Moreover, use of the paradigm concept suggests, and Dosi's remarks confirm, that technological progress must be relativised to some paradigm and makes little sense as a transparadigmatic measure of growth. To what extent does this view supply an adequate account of TC?

There is clearly a restricted sense in which one can speak of technical progress as confined to a particular technology or family of technologies. We can certainly speak of advances in propeller-driven aircraft design or thermionic valve technology long after the advent of the turbojet or the semiconductor technologies. It is equally clear that certain technological problems are specific to a particular technology and do not even arise in the context of a 'competitor'. The problems of attaining supersonic speeds or stratospheric heights, for instance, do not arise in the context of the prop-driven aircraft technology. But the fact that different technologies may reasonably be said to 'compete' with each other suggests not only that they do share many 'problems' but that they also possess many common technical variables and economic constraints. It seems hardly possible to explain TC unless this common ground is identified and analysed.

Consider Dosi's own example of the semiconductor industry:

> how could it have been possible to compare in the 1950s the thermionic valve technology with the emerging semiconductor technology? Even *ex post* (i.e., now) when most of the common dimensions (e.g., size and density, speed, costs, energy consumption, etc.) show the striking superiority of the semiconductor technology, valves still maintain in some narrow technological dimensions their advantage. (*ibid*. p. 159)

Here Dosi seems to overlook a basic feature of his own model, namely that even within a given paradigm, technological progress is (at least two-ways) relativised. Since the progress of even a single technology is a matter of trade-offs between different variables or 'dimensions', the only essential difference when we come to cross-technological comparisons is that not all relevant dimensions need be shared by each technology. But this only emphasises once again that technological progress is rarely *cumulative* in the fullest sense of the term. In some cases, it may however be 'very nearly' cumulative across a wide range of shared dimensions, as Dosi's own example illustrates.[15]

When a new technology emerges it often 'promises' more than it actually 'delivers'. When analysed in purely paradigm-bound terms, therefore, even the rate of progress of the new technology may be less than that of an established forerunner. But this point of view deprives us of the means to explain the success of the new technology. For this we need to note that a new technology is often accompanied by new and tougher perceived standards (performance, cost, safety, etc.), and these standards, whether potential or realised, are as a rule comparable with existing measures already in use, indeed they will usually be 'derived' from them. The point, then, is that whilst an established technology may continue to make 'good progress' by its own, internal standards, it may look quite regressive, even obsolete, from the vantage point of a newcomer. To return to our earlier example of aircraft design, a problem like that of achieving supersonic speeds could hardly have been conceived as an open problem before the advent of the turbo-jet engine, but once perceived and 'solved' by the new technology it might well be regarded as an 'anomaly' in the context of the earlier prop-driven tradition of engine design. In general, then, comparisons of different technologies do seem to be available and are, as a matter of routine, actually carried out. They are relevant, moreover, not only for explaining particular technological 'choices', but also for arriving at any adequate account of the nature and determinants of technological change.

6. THE PRINCIPLE OF TECHNOLOGICAL-INDUSTRIAL CONTINUITY

Giving up the idea of technological paradigms (and associated features of the Kuhnian model) would not necessarily detract from many of the positive contributions made by approaches like Dosi's. It would, however, open the way to viewing TC as a much more incremental and continuous process than the Kuhnian picture would suggest.

There are, moreover, independent reasons for thinking that the 'continuous' rather than the 'discrete' or 'disruptive' image of TC may be the more adequate one. We shall conclude by considering briefly two arguments that seem to support this view.

The first, which we touched on earlier, concerns the nature of technological knowledge as opposed to scientific knowledge. The former is distinguished from the latter in the first instance by its *practical* orientation, and this feature supports the contention that technological *knowledge* is basically cumulative, even if technological *progress* (like scientific progress) need not be. The second line of argument concerns the specific relation of technology to industrial structures and markets. Since technology must be conceived as an integral part of the industrial and economic framework, TC is in large measure constrained and governed by the evolution of that framework as well as being itself a determining factor of economic change. Since a technology is more than just a fragment of applied science, it cannot be analysed as something 'divorced from' the industrial structure in which it is grounded. The effect of this structure on TC is clearly a many-sided and complex matter, but one of its marked characteristics might be summed up by the label 'conservativeness'. Entrepreneurs are, on the whole, 'conservative' in their aversion to economic risk-taking. But there is another sense of 'conservativeness' that also applies more directly to the industrial structure: it is, very roughly, the need to accommodate technical innovations within an existing industrial framework (technical, manufacturing, marketing, etc.). Naturally, since introducing innovations may require adjustments to this system and may bring about further changes in it, whether industrial or economic, the structure is not strictly a fixed 'datum' here, but rather an evolving, dynamic entity. Nevertheless, the changes will tend to be gradual and piecemeal, and one will tend to select a new technological option by trading-off the need or desire to innovate with the ability to adapt technically and economically to the new process or product.

An example may help to illustrate the point. In the United States, the early phases of the development of nuclear reactors for civilian use were characterised by a pluralistic approach in which various different types of reactor technology were simultaneously explored. In 1954, in the USA, the Joint Committee on Atomic Energy concluded that of the five different reactor technologies currently under development, the pressurised water reactor (PWR) appeared to be the least promising due to its conservative design and poor long-term prospects.[16] Yet within five years the light water reactor had emerged as the clear front-runner, the Westinghouse (PWR) and General Electric (BWR) technologies began to establish themselves on the world market and were to set the 'standards' for nuclear power generation over the next three decades.

The success of the LWR was clearly due to a combination of technical and commercial factors. Being in many respects less 'advanced' than other technologies, the LWR initially presented fewer obstacles in being scaled-up from the prototype stage, so that it offered the possibility for a vitally early demonstration of the commercial potentials of nuclear power. More important, however, is that GE and Westinghouse opted for the LWR technology because of its close affinities with pre-existing knowledge and experience in the field of conventional power generation.[17] To that extent it is important to distinguish different stages in the development of a technology. In the early stage of applied research and development, including work on prototypes and experimental systems, the limitations of the current industrial structure are scarcely relevant: one enjoys a certain liberty to try out new experimental solutions, techniques and materials that may have little in common with experience previously acquired by the industry. In the subsequent phases leading to commercialisation, on the other hand, the influence of the industry, as the eventual locus of plant and component manufacture, is bound to make itself felt.

And here the situation may change. Consider for example the scaling-up of prototype reactors, a process that may involve changes in components and materials, as well as performance. The problems encountered here are certainly no less crucial than those surrounding the construction of prototypes in the first place. Consequently, if commercial viability is to be the target, the industry is bound at this point to prefer, where possible, technological solutions that can be mastered within its existing structure, or a structure that demands the least radical alteration. In the US such a choice in selecting a technology to be pursued into the commercialisation stage encouraged autonomous technological advances for those options which could be developed whilst maintaining a certain continuity with the pre-existing industrial structure. Other, more "innovative" paths were to stay at the experimental level and be undertaken under governmental support, not directly by the industry. And they have remained "experimental" till today.

Examples of a similar kind abound in many fields of technological growth, in the aerospace industry, in tele-communications, in micro-electronics, and so forth. Collectively, what they illustrate might be tentatively formulated as a general principle governing technological change: let us call it *the principle of technological-industrial continuity*. What the principle underlines is that whilst economic conditions of high rates of growth of demand, together with technological opportunity, may act as a stimulus to innovate, the industrial and institutional structures also exert a kind of 'pullback' effect, checking the rate and constraining the direction of technical change. A new invention or a design patent may mark an advance with respect to some purely 'technical' dimensions. But prototypes and drawing-board plans are not innovations, even if they enlarge the store of technical knowledge. Technological growth is additionally a matter of manufacturing and marketing. And, since industrial (and market) structures are not instantly and freely 'adjustable', innovations and technological changes are in a certain sense 'tailored' to their fit.

As a consequence, terms like 'radical innovation' need to be used with a measure of caution. An innovation seldom represents a complete break with the past, but is rather a suitable mix of the old and the new.[18] Thus, a technological option that to a greater extent draws on available resources, appeals to familiar techniques, and can be accommodated within existing institutional frameworks and industrial and market structures, may have the better chances to succeed. This observation applies not only to development within an established technology, but also to periods of technological transition, as the example of nuclear power indicates. What might appear as a rapid and 'disruptive' change, therefore, may turn out on closer analysis to be gradual and incremental.

For this reason, concepts like 'extraordinary technology' and 'technological revolution', if used by analogy with science, seem to be misnomers. Applied to scientific change, the metaphor of 'revolution' can be useful in underscoring the radical changes of conceptual framework that occasionally occur in the development of a scientific discipline. It indicates that a quite different way of conceptualising a certain domain of phenomena has won acceptance. In the final resort, the epistemic utilities of science are not rigidily attached to any particular conceptual scheme, and, in overthrowing a given theory, we may happen to discard the conceptual scheme that goes with it. If technology possesses anything analogous to a conceptual scheme, it is certainly quite unlike the 'languages' and 'models' associated with science. If, in technology studies, the idea of a conceptual scheme makes sense at all, it has to be characterised as a complex system, comprising governmental, institutional and legal frameworks, scientific, industrial and economic structures, together with their many-sided interrelations. This is the 'system' in which technological change takes place. It is a system in flux, but, at least in modern industrialised societies, it is not a system that can be abandoned wholesale or replaced at a stroke.

NOTES

[1] Such a distinction is not often made in the literature. This carries no great consequences when the discussion is on the broad level. However misunderstandings can often be avoided when the distinction is properly drawn.

[2] For attempts to develop 'Kuhnian' accounts of TC, see e.g., Wojick (1979), Constant (1980) and Dosi (1982). Wojick's and Constant's views are critically examined by Gutting (1984); Dosi's approach will be treated in Section 5 below.

[3] Cf. for example N. Kaldor (1954), p. 53 'The Relation of Economic Growth and Cyclical Fluctuations', *The Economic Journal* **64**, 53–71.

[4] Galbraith, as one of the most prominent of Schumpeter's followers, emphasised further the importance of the large, oligopolistic firms as the locus of innovation, and stressed the supremacy of technical competition over price competition. Empirical studies have focussed on testing the positive correlation between monopolistic proxies like size of the firm and concentration, and R&D intensity and innovation rates.

[5] Criticism was exercised from the very beginning on the long range time series and on the dubious estimation of capital. It was argued that to measure the increase in global productivity as a function of technical progress had no real meaning. However, major attacks were directed at the concept of an aggregate production function and on the unrealistic neoclassical assumptions, which gave rise to one of the most long-lived debates of economic theory: the 'two Cambridge' controversy. Further criticism pointed out that the increase in product attributable to the increase in production factors depends on the formation of the production function, so that if one takes a homogeneous PF of first degree, the increase in product deriving from the increase in factors is smaller, ceteris paribus, than what it would be with a PF of a degree larger than one. Later criticism was addressed to the feasibility of the method itself.

[6] There are at least 15 different definitions of neutral TP. The most famous, however, are Hicks', Solow's and Harrod's. Hicks' and Harrod's models have aroused considerable debate, and opinions are split over whether the former is perhaps better adapted to

analyse exogenous TP, and the latter better suited to analyse TP dependent on endogenous factors, especially at a macrolevel. For more details on Hicks' neutrality, cf. Steedmann, I., On the impossibility of Hicks neutral technical change, *The Economic Journal* **95**, 746–758.

[7] This approach moves from a general production function, the model CES (constant elasticity substitution) of the type

$$Y = \gamma[\delta K^{-\rho} + (1-\delta)L^{-\rho}]^{-1/\rho},$$

where ρ is the substitution parameter, δ the distribution parameter, and γ is the efficiency parameter. The properties of the CES are first degree homogeneity and constancy of substitution elasticity between K and L. In this case, both Cobb–Douglas and Solow's PFs characterised by complementarity and substitutibility among factors are a particular case ($\rho = 0$) of the CES function. Endogenous technological progress is achieved when efficiency parameters increase; when, with a capital intensity >1, the isoquants turn towards the capital axis (the distribution parameter increases); when, with a capital intensity <1, the substitution elasticity increases (the isoquants' slope diminishes); and when the homogeneity parameter increases (the distance between the isoquants is reduced).

[8] However, the Keynesian critics were also not exempt from criticism. As Blaug (1963), p. 131, puts it: "The neoclassical idea of a given state of knowledge is admittedly an abstract one. But the concept of a given rate of change of knowledge is almost metaphysical". For reason of space no account is given here of the post-Keynesian analysis. For reference, see among others Stoneman (1983).

[9] Cf. Nelson and Winter (1974), p. 890.

[10] For a comprehensive account of the research and theory in this field, see, among others Kamien and Schwartz (1982).

[11] Much of the empirical research of the '60s and '70s has aimed at testing the relationship between innovation (usually given by proxies like R&D indicators) and firms' performance, and has assumed the hypothesis of R&D as an input stably related to output (patents). This stability was held to form the basis of a virtuous circle in which innovations determine a rise in the growth rate of the innovating firms and lead to an increase in productivity and hence profitability.

[12] Though somewhat simplified, from Nelson and Winter's computer simulations it seems to emerge that innovation pays dividends in cases of restrained competition, whilst imitation seems to be the best behavioural rule under the conditions of aggressive competition.

[13] For a case-study of the invention of radio, see Aitken (1978). As Aitken's account plainly shows, the successful implementation of radio technology cannot be attributed to any particular 'failures' previously perceived in cable methods of message transmission (nor explained by a demand-pull of the market). The emergence of the new technology, in Aitken's words, "called for imagination and salesmanship. It called for the kind of entrepreneurship that could create needs, not merely serve them" (p. 96).

[14] For an elaboration of this view of science, see especially Laudan (1977).

[15] Some arguments against the appropriateness of a 'Kuhnian' account of the development of the semiconductor technology are given in Schopman (1981).

[16] The other four reactor types were an experimental boiling water reactor (EBWR), a

thermal reactor cooled with sodium (SRE), a homogeneous reactor with water-uranyl sulphate solution (HRE), and a fast sodium reactor (EBR). For details see, e.g., Nehert (1966).

[17] This point is elaborated in greater detail in Pearce (1986).

[18] We can agree here with Sahal (1981, p. 37): "This is not to say that radical advances in technology do not occur. Rather, the major innovations are made possible by numerous minor innovations. The cumulative impact of many seemingly minor changes in technology often tends to be quite substantial".

REFERENCES

Aitken, H.: 1978, 'Science, Technology and Economics: The Invention of Radio as a Case Study', in W. Krohn, E. Layton and P. Weingart (eds.), *The Dynamics of Science and Technology*, D. Reidel, Dordrecht.

Blaug, M.: 1963, 'A Survey of the Theory of Process-Innovations', *Economica* **30**, 13–32.

Constant, E.: 1980, *The Origins of the Turbojet Revolution*, John Hopkins University Press, Baltimore.

Dosi, G.: 1982, 'Technological Paradigms and Technological Trajectories. A Suggested Interpretation of the Determinants and Directions of Technical Change', *Research Policy* **11**, 147–62.

Dosi, G.: 1984, *Technical Change and Industrial Transformation*, Macmillan, London.

Freeman, C.: 1977, 'Economics of Research and Development', in I. Spiegel-Roesing and D. De Solla Price (eds.), *Science, Technology and Society*, Sage Publ., London.

Gutting, G.: 1984, 'Paradigms, Revolution, and Technology', in Laudan (1984).

Kamien, M. and Schwartz, N.: 1982, *Market Structure and Innovation*, Cambridge University Press, Cambridge.

Laudan, L.: 1977, *Progress and its Problems*, RKP, London.

Laudan, R. (ed.): 1984, *The Nature of Technological Knowledge. Are Models of Scientific Change Relevant?*, D. Reidel, Dordrecht.

Nehert, L. C.: 1966, *International Marketing of Nuclear Power Plants*, Indiana University Press, Bloomington.

Nelson, R. and S. Winter: 1974, 'Neoclassical vs. Evolutionary Theories of Economic Growth', *Economic Journal* **84**, 886–905.

Nelson, R. and S. Winter: 1977, 'In Search of a Useful Theory of Innovation', *Research Policy* **6**, 36–76.

Nelson, R. and S. Winter: 1982, *An Evolutionary Theory of Economic Change*, Belknap, Cambridge.

Pearce, M. R.: 1986, *Technology, Competition and State Intervention*, D. Phil. Thesis, Sussex.

Rosenberg, N.: 1976, *Perspectives on Technology*, Cambridge University Press, Cambridge.

Sahal, D.: 1981, *Patterns of Technological Innovation*, Addison-Wesley, Reading, Mass.

Schopman, J.: 1981, 'The History of Semiconductor Electronics – A Kuhnian Story?', *Zeitschrift für allgemeine Wissenschaftstheorie* **12**, 297–302.

Solow, R.: 1957, 'Technical Change and the Aggregate Production Function', *Review of Economics and Statistics* **39**, 312–20.
Stoneman, P.: 1983, *The Economic Analysis of Technological Change*, Oxford University Press, Oxford.
Wojick, D.: 1979, 'The Structure of Technological Revolutions', in G. Bugliarello and D. Doner (eds.), *The History and Philosophy of Technology*, University of Illinois Press, Urbana.

Manuscript received 25 January 1988

Maria Rosaria Di Nucci Pearce
Institut für Volkswirtschaftslehre
Technische Universität Berlin

David Pearce
Institut für Philosophie
Freie Universität Berlin

W. BALZER AND E.-W. HAENDLER

ORDINARY LEAST SQUARES AS A METHOD OF MEASUREMENT*

Statistical estimation plays a decisive role within the empirically oriented branches of economics and various other social sciences. Statistical methods are also applied in the natural sciences; the empiricity of the natural sciences is, however, anchored in a more direct mode in "repeatable" measurement, which assigns statistical estimation only a secondary status.

On a narrow conception, measurement consists in "comparison with a unit", a conception which, together with the paradigm of fundamental measurement, has been widely accepted in psychology and the social sciences, too. Yet typical examples of "measurement" in the natural sciences can be subsumed under this concept only at the expense of severe biasedness. Typically, what we encounter here are situations in which theoretical equations, "theories", are employed in order to "calculate" the desired "measured" values. We will adopt here without argument a broad conception of measurement according to which even such theory-dependent methods of determination are termed measurement: the structuralist view of measurement.[1] According to this view the calculation of parameters from a set of equations, for instance, constitutes a method of measurement for these parameters, provided there is a unique solution.

The latter procedure is of course analogous to an estimation of parameters in the social sciences. So, is estimation a method of measurement (in the broad, structuralist sense)? This is not a mere question of terminology. Rather, the attempt of answering this question reveals precise distinctions between related procedures (which we call measurement) in "statistical" social science on the one hand and in the natural sciences on the other. Besides, our investigation will shed light on some fundamental methodological differences between the natural and the social sciences.

We will restrict our analysis to the simple case of applying the method of ordinary least squares (OLS) in order to estimate the two parameters of a linear demand function, a case typical for a wide range of similar elementary applications of OLS. Our aim is to

Erkenntnis **30** (1989) 129–146.

subsume OLS under the general structuralist concept of a method of measurement. By this we do not want to adopt once again the imperalist strategy of recommending methodological ideas from physical science to the social scientist. It will turn out (in Section V) that the notion of theory-dependent measurement which was developed to cover the respective phenomena within physical science has to be generalized to be applicable to OLS. We introduce the notion of a regression method of measurement which covers methods like OLS as well as "ordinary" theory-dependent measurement in the natural sciences (and, of course, fundamental measurement as well). This enables us to work out features common to both the natural and the social sciences as well as to illustrate fundamental differences between the natural and the social sciences.

Furthermore, we discuss the question of how to justify OLS as a method of measurement (Section IV). In this context, again, we encounter strong similarities but at the same time clear differences between the natural and the social sciences.

I. MEASUREMENT

The structuralist view of measurement starts from considering actual procedures of measurement and focuses on the structure of single, isolated processes of measurement, in the course of which one value, the *measured value*, is produced. Such procedures involve a real system, the measuring apparatus, and in most cases some theoretical equations which come from one single or from several established theories and which are used in order to calculate the measured value. The real system thus is represented as a model or a chain of models of one or several theories which *govern* the process of measurement. Such models we call measuring models. A measuring model therefore represents one single process of measurement (or even only a part of it) as far as it can be subsumed under some given theory. As a borderline case the theory may be a mere theory of measurement, like in fundamental measurement the "theory" of extensive systems.[2]

Measuring models have at least four general features in common. First, they are characterized by a law-like proposition (the laws of the theory governing the process of measurement, see (D1–3) below). Second, the measured value in each measuring model is uniquely determined by other parts of the model, and by the law characterizing the model (D1–5). Third, the measured value is effectively computable from other "parts" of the model (D1–6), and fourth, the measured value is a continuous function of those other parts from which it can be computed (D1–7).[3]

The law-like propositions may be of various kinds. They may be proper axioms of established theories, or simpler equations "derived" from more complicated such axioms. In fundamental measurement they are the axioms put forward by measurement theorists. Continuity is required in order to exclude contrived cases, like setting the measured value equal to 1 by definition. In such cases there is no point in measuring since the "measured value" is fixed purely conceptually. Usually, the "same" measuring model will be used to measure not just one value but a set of such values (by repeated application). Accordingly, uniqueness may be required not just for one value but for a set of values of the function F to be measured. For reasons of simplicity we even require that F be uniquely determined for all of its arguments.[4]

These conditions may be integrated into a general definition of a measuring model as follows.

(D1) x is a *measuring model for function F characterized by B, Σ,* τ and $\approx$ iff there exist $D_1, \ldots, D_k, A_1, \ldots, A_m, R_1, \ldots, R_n$ such that

(1) τ is a type and $x = \langle D_1, \ldots, D_k, A_1, \ldots, A_m, R_1, \ldots, R_n, F\rangle$ is a set theoretic structure of type τ

(2) F is a function

(3) B is a law-like statement, and valid in x

(4) $\approx$ is an equivalence relation on the class of all functions of the type of F

(5) F is uniquely determined by B in x (up to $\approx$)

(6) each function value $F(a)$ can be computed from a finite substructure of $\langle D_1, \ldots, R_n\rangle$ (up to $\approx$, and after appropriate encoding)

(7) $\Sigma = \langle \Sigma_1, \Sigma_2\rangle$ is a pair of topologies such that B defines a function $\langle D_1, \ldots, R_n\rangle \rightarrow F$ which is piecewise continuous w.r.t. Σ_1 and Σ_2.

The values of F we call *measured values*. The special case in which an explicit definition D for F in terms of the other components $D_1, \ldots, R_n$ is "contained" in statement B will be of particular importance in the following. In this case we speak of measuring models for the defined term F, by which label we refer to structures from which F has been removed ((D2-2) below).

(D2) x is a *measuring model for the defined term F characterized by* B, Σ, τ, $\approx$ and D iff there exist $D_1, \ldots, D_k$, $A_1, \ldots, A_m$, $R_1, \ldots, R_n$ such that
(1) $\langle D_1, \ldots, R_n, F\rangle$ is a measuring model for function F characterized by B, Σ, τ and $\approx$,
(2) $x = \langle D_1, \ldots, R_n\rangle$,
(3) D is an explicit definition of F in terms of x,
(4) there is a law-like statement B^* such that $B(D_1, \ldots, R_n, F)$ is equivalent with $B^*(x) \wedge D(x, F)$.

The law-like proposition B (or B^*) of course can be extracted only from the investigation of a whole class of many similar measuring models. Such a class we call a method of measurement. The connection to the usual use of the word "method" is established by observing that each method determines the class of all systems in which it can be successfully applied, and conversely.

(D3) M_m is a *method of measurement for function* $\{F\}$ iff there exist B, Σ, τ and $\approx$ such that M_m is the class of all measuring models for function F characterized by B, Σ, τ and $\approx$, and $M_m \neq \emptyset$.

(D4) M_m is a *method of measurement for the defined term* $\{F\}$ iff there exist B, Σ, τ, $\approx$ and D such that M_m is the class of all measuring models for the defined term F characterized by B, Σ, τ, $\approx$ and D, and $M_m \neq \emptyset$.

In the natural sciences there are many cases of measurement in the course of which a given theory is used and presupposed in order to calculate the measured values. Any method of measurement with this feature we call theory-dependent.

(D5) If T is a theory[5] with class M of models then M_m is a *T-dependent* method of measurement iff, for some $\{F\}$, M_m is a method of measurement for $\{F\}$ or M_m is a method of measurement for the defined term $\{F\}$, and $M_m \subseteq M$.

As a paradigm for theory-dependent measurement let us consider the measurement of mass by collisions as governed by classical collision mechanics (CCM). The models (M(CCM)) are defined as follows. $x \in M$(CCM) iff x has the form $\langle P, \{b, a\}, \mathbb{R}, v, m\rangle$ and (1) P is a non-empty, finite set (of "particles"), (2) $\{b, a\}$ is a two-element set (of "instants": "*b*efore" and "*a*fter" the collision), (3) $m: P \to \mathbb{R}^+$ ("mass-function"), (4) $v: P \times \{b, a\} \to \mathbb{R}^3$ ("velocity-function"), (5) $\Sigma_{p \in P}\, m(p)v(b, p) = \Sigma_{p \in P}\, m(p)v(a, p)$ (law of conservation of total momentum).

The models are intended to describe collisions of two or more particles. Let $\bar{m}$ denote the term "mass". A method of measurement M_m(CCM) for $\bar{m}$ is defined as follows. $x \in M_m$(CCM) iff (1) $x = \langle P, \{b, a\}, \mathbb{R}, v, m\rangle \in M$(CCM), (2) P is a two-element set ($P = \{p, p'\}$), (3) v is such that all $v(p^*, t)$, $p^* \in P$, $t \in \{b, a\}$ are on a straight line, (4) for $t \in \{b, a\}$, $v(p, t)$ and $v(p', t)$ have opposite direction, (5) $v(p, b) \neq v(p', a)$.

The topologies required in (D1–7) are generated by neighbourhoods defined as follows. For $x = \langle P, \{b, a\}, \mathbb{R}, v\rangle$, $y \in U_x^\epsilon$ iff $y = \langle P, \{b, a\}, \mathbb{R}, v'\rangle$ and for all $p \in P$ and $t \in \{b, a\}$, $|v(p, t) - v'(p, t)| < \epsilon$. For $m: P \to \mathbb{R}^+$, $m' \in U_m^\epsilon$ iff $m': P \to \mathbb{R}^+$ and for all $p \in P$: $|m(p) - m'(p)| < \epsilon$. $\approx$ is given by: $m \approx m'$ iff $\mathrm{Dom}(m) = \mathrm{Dom}(m')$ and there exists $\alpha \in \mathbb{R}^+$ such that for all $p \in \mathrm{Dom}(m)$: $m(p) = \alpha \cdot m'(p)$. The value $m(p')$ in $x \in M_m$(CCM) is computable from v up to some $\alpha \in \mathbb{R}^+$, and, for given such α, it varies smoothly with variation of v. It is not difficult to prove that M_m(CCM), in fact, is a method of measurement for $\{m\}$.

As an example of a method of measurement for a defined term F think of the method of determining the mean velocity of a uniformly moving particle. If $s_p(t)$ indicates the position of particle p at time t, and if p moves uniformly, then its mean velocity v_p (which incidentally is also its actual velocity) is defined as

$$v_p = \frac{s_p(t) - s_p(t')}{t - t'}$$

where t, t' are different instants (which we treat as real numbers for the sake of simplicity) and $t' < t$. A corresponding measuring model for v_p has the form $\langle\{p\}, T, \mathbb{R}^3, s_p\rangle$ where $T \subseteq \mathbb{R}$ is an open interval, s_p: $T \to \mathbb{R}^3$ is such that its image is contained in a straight line and Ds_p is a constant function, and v_p is defined as just indicated. It is easy to define the corresponding method of measurement for the defined term $\{v_p\}$.

II. OLS: A RECONSTRUCTION OF LINEAR STOCHASTIC DEMAND SYSTEMS

We now turn to our example from applied econometrics. A linear demand system is a system in which the purchases, d_t, for a commodity (expressed in quantitative, numerical form) are supposed to be a linear function of the price, p_t, of this commodity, where both prices and purchases may change over time, t.

It must be emphasized that the existence of a linear relationship between observed purchases and observed prices does not provide much justification for speaking of "demands" and "demand systems". Our focus in this paper is, however, not on the criteria of identity for the function to be measured. We investigate parameter estimation by means of the method of ordinary least squares on the basis of the very elementary example of a linear relationship between observed purchases and observed prices. The result of our inquiry applies likewise to the more intricate demand theories which actually represent the state of the art.

While theoretical as well as practical considerations may suggest a linear relationship between prices and purchases, the observed time-series $\langle p_t, d_t\rangle$, $t = 1, \ldots, n$ usually will not satisfy a linear equation

$$d_t = \beta_1 + \beta_2 p_t \qquad \text{for all } t \leq n.$$

In order to accomodate for deviations, p_t and d_t are regarded as values of random variables $\tilde{p}_t$, $\tilde{d}_t$, both defined on some suitable set W of events (which ordinarily is not made explicit), and a disturbance variable $\tilde{u}_t$ is introduced which accounts for all those influences on the purchases which are not allowed for by the prices (e.g., the influence of the temperature on the purchases of sodas). These hypothetical random variables are assumed to form the following linear relationship:

$$(1) \qquad \tilde{d}_t(w) = \beta_1 + \beta_2 \tilde{p}_t(w) + \tilde{u}_t(w), \text{ for all } t \leq n \text{ and } w \in W.$$

(See (D7–3) below). p_t and d_t are taken as particular realizations of $\tilde{p}_t$, $\tilde{d}_t$: $\tilde{p}_t(w_o) = p_t$ and $\tilde{d}_t(w_0) = d_t$ for some $w_o \in W$ where w_o is the event which actually has occurred (D7–2).

(D6) x is a *potential linear stochastic demand system* ($x \in M_p(\text{LSD})$) iff there exist $n, p, d, W, \mathfrak{A}, \mu, \tilde{p}, \tilde{d}, \tilde{u}$ such that $x = \langle n, p, d, W, \mathfrak{A}, \mu, \tilde{p}, \tilde{d}, \tilde{u}, \beta \rangle$ and
(1) $n \in \mathbb{N}$, $n > 0$,
(2) $p, d \in \mathbf{R}^n_+$,
(3) $\langle W, \mathfrak{A}, \mu \rangle$ is a probability space,
(4) $\tilde{p}, \tilde{d}, \tilde{u}: W \to \mathbf{R}^n$ are random variables,
(5) $\beta \in \mathbf{R}^2$.

For $\tilde{p}, \tilde{d}, \tilde{u}$ and $t \leq n$ we define functions $\tilde{p}_t: W \to \mathbf{R}$ by $\tilde{p}_t(w) = (\tilde{p}(w))_t$ etc.

(D7) x is a *linear stochastic demand system* ($x \in M(\text{LSD})$) iff there exist $n, p, d, W, \mathfrak{A}, \mu, \tilde{p}, \tilde{d}, \tilde{u}, \beta$ such that $x = \langle n, p, d, W, \mathfrak{A}, \mu, \tilde{p}, \tilde{d}, \tilde{u}, \beta \rangle$ and
(1) $x \in M_p(\text{LSD})$,
(2) there exist $V \in \mathfrak{A}$ and $w_o \in V$ such that $\mu(V) > 0$, $\tilde{p}(w_o) = p$ and $\tilde{d}(w_o) = d$,
(3) for all $w \in W$ and $t \leq n$: $\tilde{d}_t(w) = \beta_1 + \beta_2 \tilde{p}_t(w) + \tilde{u}_t(w)$,
(4) for all $t \leq n$ and all $a \in \text{Rge}(\tilde{p})$: the conditional expectation of $\tilde{u}_t$ under condition a, $E(\tilde{u}_t \mid a)$, is zero.

Some auxiliary definitions are needed for the following.

(D8) (a) If $a = \langle a_1, \ldots, a_n \rangle$ and $b = \langle b_1, \ldots, b_n \rangle \in \mathbf{R}^n$, and $f = \langle f_1, \ldots, f_n \rangle$ and $g = \langle g_1, \ldots, g_n \rangle$ are functions, $f, g: W \to \mathbf{R}^n$, then

$$\bar{a} = \frac{1}{n} \sum_{j \leq n} a_j, \qquad s^2(a) = \frac{1}{n} \sum_{j \leq n} (a_j - \bar{a})^2,$$

$$\text{cov}(a, b) = \frac{1}{n} \sum_{j \leq n} (a_j - \bar{a})(b_j - \bar{b}).$$

$\bar{f}$, $s^2(f)$, $\text{cov}(f, g): W \to \mathbf{R}^n$ are defined by $\bar{f}(w) = \overline{f(w)}$, $s^2(f)(w) = s^2(f(w))$, $\text{cov}(f, g)(w) = \text{cov}(f(w), g(w))$

(b) If $x = \langle n, p, d, W, \mathfrak{A}, \mu, \tilde{p}, \tilde{d}, \tilde{u}, \beta \rangle \in M_p(\text{LSD})$ we write

$$\hat{\beta}_2(p, d) = \frac{\text{cov}(p, d)}{s^2(p)}, \quad \hat{\beta}_1(p, d) = \bar{d} - \hat{\beta}_2(p, d) \cdot \bar{p}$$

and, for $w \in W$:

$$\tilde{\beta}_2(w) = \frac{\mathrm{cov}(\tilde{p}, \tilde{d})(w)}{s^2(\tilde{p})(w)}, \quad \tilde{\beta}_1(w) = \bar{\tilde{d}}(w) - \tilde{\beta}_2(w) \cdot \bar{\tilde{p}}(w).$$

The method of measurement we want to consider is the following. By minimizing the squared deviations of the observed data from a hypothetical straight line, OLS recommends the values $\hat{\beta}_1(p, d)$, $\hat{\beta}_2(p, d)$ as defined in (D8–b) above as estimations for the unknown parameters β_1, β_2 which occur in the models (see (D7–3)). It is common to call these unknown, hypothetical parameters β_1, β_2 the *true* parameters. "True parameter" is thus defined as "parameter which occur(s) in (one of) the model(s) which (are) is assumed to capture the real system under study." We adopt this usage without inquiring into its possible philosophical interpretations.

The choice of $\hat{\beta}_i$ as candidates for the true values β_i is justified by the fact that the suggested values $\hat{\beta}_i$ are instances of *estimators* $\tilde{\beta}_i$ as defined in (D8–b) above, which exhibit some desirable statistical properties. In the first place $\tilde{\beta}_i$ are *unbiased*. This means that $\hat{\beta}_i(p, d)$ are realizations of $\tilde{\beta}_i$ (for some $w_o \in W$, $\tilde{\beta}_i(w_o) = \hat{\beta}_i(p, d)$), and the mean value of $\tilde{\beta}_i$, i.e., the integral of $\tilde{\beta}_i$ over W with respect to μ, is identical with β_i, i.e., with the "true" but unknown value hypothesized to make equations (1) true.

Unbiasedness cannot, of course, be proved right away. Further idealizing mathematical assumptions have to be postulated. A classical set of idealizing assumptions is given in the following definition.[6]

(D9) x is a *classical* linear stochastic demand system ($x \in$ CLSD) iff there exist $n, p, d, W, \mathfrak{A}, \mu, \tilde{p}, \tilde{d}, \tilde{u}, \beta$ such that $x = \langle n, p, d, W, \mathfrak{A}, \mu, \tilde{p}, \tilde{d}, \tilde{u}, \beta \rangle$ and

(1) $x \in M(\mathrm{LSD})$,
(2) for all $t, t' \leq n$, $t \neq t'$ and all $a \in \mathrm{Range}\ (\tilde{p})$:
  (2.1) *the conditional expectation of* $u_t u_{t'}$ under the condition that $\tilde{p}$ takes value a is zero, i.e., $E(\tilde{u}_t \tilde{u}_{t'} \mid a) = 0$,
  (2.2) the conditional variance of $\tilde{u}_t$ under the condition that $\tilde{p}$ takes value a exists and does not vary with t, i.e., there exists σ^2 such that $V(\tilde{u}_t \mid a) = \sigma^2$,
  (2.3) $\tilde{u}_t$ is normally distributed,
  (2.4) $V^* := \{w \in W / s^2(\tilde{p})(w) > 0\} \in \mathfrak{A}$ and $\mu(V^*) = 1$.

(T1) If $x = \langle n, p, d, \ldots, \beta \rangle$ is a CLSD then
(a) for $i = 1, 2$: β_i equals the expectation of $\tilde{\beta}_i$, i.e.,

$$\beta_i = E(\tilde{\beta}_i) = \int_W \tilde{\beta}_i \, d\mu,$$

(b) for each $w_o \in W$ such that $\tilde{p}(w_o) = p$ and $\tilde{d}(w_o) = d$, and for $i = 1, 2$:

$$\tilde{\beta}_i(w_o) = \hat{\beta}_i(p, d),$$

Proof. For (a) see: e.g., Schneeweiß (1971) p. 59. (b) follows directly from (D8).

In the following we will concentrate on the property of unbiasedness and leave other desirable properties for statistical estimators like consistency or sufficiency out of consideration. Our investigation applies to these statistical properties as well, but at the cost of formal complication. We only note that the generalization expressed in (D12) below may be strengthened (and thereby specialized again) such that further criteria for estimators can be taken into account.

III. CLASSICAL LINEAR STOCHASTIC DEMAND SYSTEMS AS A METHOD OF MEASUREMENT

The actual procedure for the estimation of the parameters of linear stochastic demand systems is this. Observe time-series $\langle p_t, d_t \rangle$ and calculate $\hat{\beta}_1(p, d)$, $\hat{\beta}_2(p, d)$. These are the best values for β_1, β_2 you can obtain. Full stop.

The rest of the story is justification. If for a moment we do not care about the justification this procedure gives rise to a method of measurement as defined in Section I.

(D10) x is a *measuring model for* the defined term $\hat{\beta}$ *by regression* ($x \in M_m(\text{CLSD})$) iff x has the form $\langle n, p, d, W, \mathfrak{A}, \mu, \tilde{p}, \tilde{d}, \tilde{u}, \beta \rangle$ where $\langle n, \ldots, \beta \rangle \in \text{CLSD}$ and $\hat{\beta} = \langle \hat{\beta}_1(p, d), \hat{\beta}_2(p, d) \rangle$

The measured values in such measuring models are the parameters $\hat{\beta}_i(p, d)$, and a corresponding method of measurement is easily defined.

(T2) $M_m(\text{CLSD})$ is a method of measurement for the defined term $\hat{\beta}$ characterized by B, Σ, τ, $\approx$ and D where

(1) D is the definition of $\hat{\beta}$ (compare (D8–b)),

(2) B is the conjunction of the axioms (D6–1) to (D6–5), (D7–2) to (D7–4), (D9–2), and (D),

(3) $\Sigma = \langle \Sigma_1, \Sigma_2 \rangle$ where Σ_1, Σ_2 are generated by neighbourhoods of the form U_z^ϵ, $U_{\hat{\beta}}^\epsilon$ defined by $z' \in U_z^\epsilon$ iff $z = \langle n, p, d, W, \mathfrak{A}, \mu, \tilde{p}, \tilde{d}, \tilde{u}, \beta \rangle$, $z' = \langle n, p', d', W, \mathfrak{A}, \mu, \tilde{p}, \tilde{d}, \tilde{u}, \beta \rangle$, $|p - p'| < \epsilon$ and $|d - d'| < \epsilon$, and $\hat{\beta}' \in U_{\hat{\beta}}^\epsilon$ iff $\hat{\beta} = \langle \hat{\beta}_1, \hat{\beta}_2 \rangle$, $\hat{\beta}' = \langle \hat{\beta}_1', \hat{\beta}_2' \rangle$ and $|\hat{\beta} - \hat{\beta}'| < \epsilon$,

(4) τ is the type of linear stochastic demand systems extended by $\hat{\beta}$,

(5) $\approx$ is identity.

Proof. $\hat{\beta}_i(p, d)$ is a continuous function up to one singularity for $p = 0$. The remaining requirements are trivially satisfied.

Note that the measuring models refer to two sets of parameters: $\langle \beta_1, \beta_2 \rangle$ which are purely hypothetical and occur in the linear equation (1), the "law" of the theory, and $\langle \hat{\beta}_1(p, d), \hat{\beta}_2(p, d) \rangle$ which are defined in terms of the data given by p and d. $\langle \beta_1, \beta_2 \rangle$ are the values one would like to determine (to measure) while $\langle \hat{\beta}_1(p, d), \hat{\beta}_2(p, d) \rangle$ are the values actually obtained. Usually, $\hat{\beta}_i(p, d)$ differs from β_i. So in what sense can we say that the method of measurement described is a method of measurement *for* β (and not only for $\hat{\beta}$, which holds trivially)?

In answering this question we refer to the statistical justification already mentioned. $M_m \in$ (CLSD) is a method of measurement *for* β inasfar as (1) the values actually measured, $\hat{\beta}_i$, are realizations of estimators $\tilde{\beta}_i$, and (2) integration over each $\tilde{\beta}_i$ yields the value β_i, i.e., the "true" value one wants to find out.

It must be emphasized that the provable identity of $\int \tilde{\beta}_i \, d\mu$ and β_i does not help to determine the desired values β_i. For of all the values of $\tilde{\beta}_i$ the only one we know is $\tilde{\beta}_i(w_o)$, where w_o is the event for which $\tilde{p}(w_o) = p$ and $\tilde{d}(w_o) = d$. These are the only data at hand. For all other $w \in W$, $w \neq w_o$, $\tilde{\beta}_i(w)$ is unknown, and therefore the value of the integral cannot be calculated.

Also it has to be noted that neither the theoretical assumption of linearity nor the true parameters β_i occurring in it are used or play any direct role in the determination of the measured values $\hat{\beta}_i(p, d)$.

IV. JUSTIFICATION

In order to discuss the justification of M_m(CLSD) and its differences to methods of measurement in the natural sciences, let us use the phrase:

> measuring model x for function F is intended to measure function G.

The reason for using this phrase is the following. Usually, if we want to measure some function G we already have at hand some, perhaps rather weak, criteria of identity for G, conditions that G has to satisfy independently of the result of the measurement process. In most cases the function we want to measure already has a name, like "mass", "utility", or "price". In these cases we do not simply perform some process of measurement and accept the result as that value we intended to measure. Rather, we have to justify why some measuring model measures price, and not, say, utility. On the other hand the measuring model actually produces *some* value, independently of whether we accept this value as adequate or not. We call these values, the function values of the function F, *measured values*.

Consider two typical cases from the natural sciences.

Case A. The function G one intends to measure is a primitive of the theory T which governs the process of measurement. In this case the criteria of identity for G are given by theory T. The function one intends to measure is the function G as determined by its role in theory T. In order to guarantee that these criteria are met in the course of measurement it is assumed that the measuring model is a proper model of theory T. If the measuring model is a model of T then the values of F actually measured satisfy rather strong conditions as given by the axioms of T: they are in this sense consistent with T. Since the criteria for G consist of these same requirements, both F and G satisfy the same criteria. In many concrete cases this entails identity of F and G.

This is the situation of theory-dependent measurement introduced in Section I. The criteria for G are given by the axioms of T, these same axioms are used and presupposed in the course of the measurement which yields values of F, and therefore the values of F are acceptable, they "are" (i.e., may be accepted as) values of the function

G one intends to measure. The situation may be depicted as in Figure 1. In this case the justification for the assertion that the method of measurement really produces the values one intends to measure is this. Because the process of measurement is assumed to be a model of the theory, the measured values of F satisfy the criteria of identity for G, the function one intends to measure. Therefore, values of F are acceptable as values for G.

Case B. The function F, whose values are the actual result of the process of measurement, is a defined term. Usually the function G, the function one intends to measure, is given by the same definition as F. Take measurement of mean velocity as an example. Mean velocity is defined in terms of positions, the measured values are calculated by using this definition, and the function one intends to measure is velocity as given by the same definition. The criteria for G will reduce

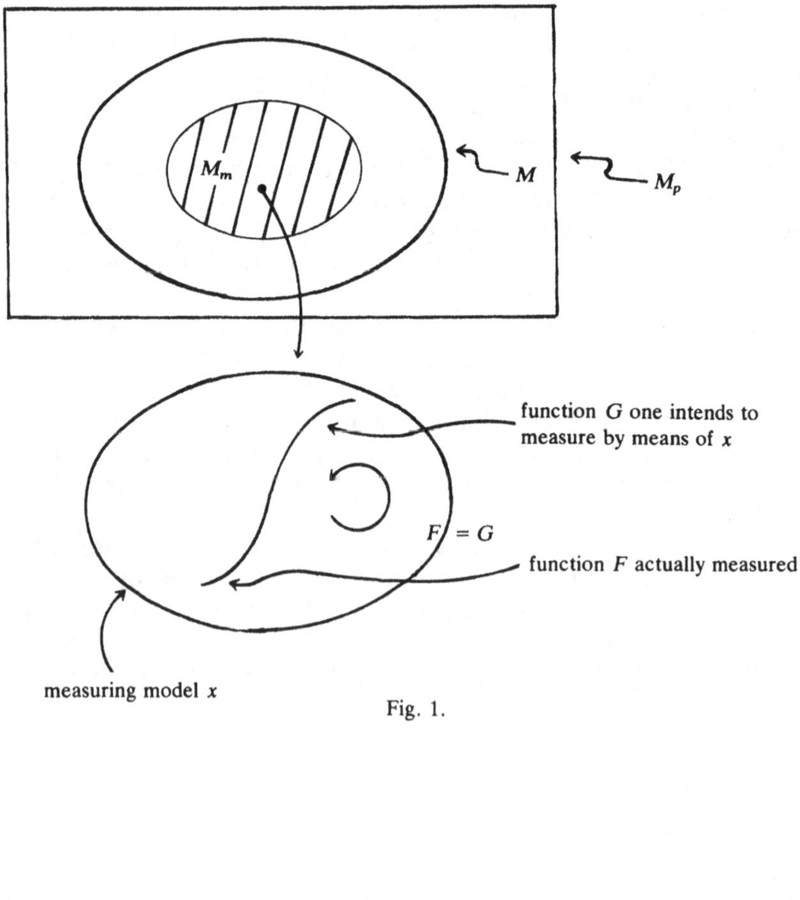

Fig. 1.

to criteria for those functions in terms of which G is defined. This leads to the situation of case A just discussed: the criteria of identity are given by the axioms of the theory governing the process of measurement. The situation is depicted in Figure 2. The important point is the following: in the natural sciences measurement of a defined term means that the function G one wants to measure by means of producing measured values of function F ***practically never*** is a primitive of the theory which is used in the respective measuring model.

What is the justification for the assertion that measurement of a defined term really produces the value one intends to measure? M_m is

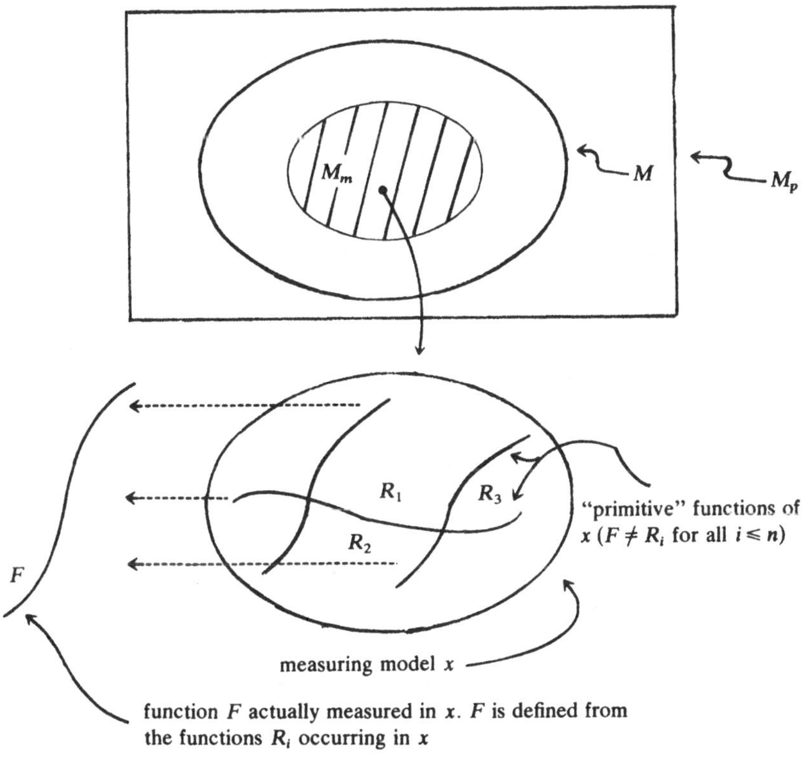

Fig. 2.

accepted as a method of measurement for G because the process of measurement is assumed to be a model of the theory, and because the definitions of F and G are identical. These assumptions entail that the *definiens* of F satisfies the criteria of identity for the *definiens* of G as given by the axioms of the theory. Therefore the *definiens* for F is acceptable as a *definiens* for G, and, since the definitions of F and G are the same, the values of F are acceptable as values for G.

We now are prepared to turn to the present example of measurement by OLS. In this case the term F actually measured, namely $\hat{\beta}$, is explicitly defined, so we are in the situation of case B just discussed. But there is a decisive difference between measurement by OLS and measurement of a defined term in the natural sciences. Now the function G one intends to measure *is* a primitive of the theory which governs the measuring model. In this case our terminology yields the following. We have measuring models for the *defined* term F which are intended to measure a *primitive* function G ("primitive" with respect to the theory governing the measuring models). The picture is as in Figure 3. In the natural sciences the measuring model is a model of some theory T, the function F actually measured is explicitly defined, and the function G one intends to measure is *not* a primitive of T. In the case of OLS the function G one intends to measure *is* a primitive of the theory.

The criteria for identity of G, i.e., for β in the case of OLS, are given by the theory from which $G(\beta)$ comes from: the "theory" of linear stochastic demand systems. These criteria will, however, usually not be met by the measured values $\hat{\beta}$. The justification for taking OLS as a method of measurement for G was worked out in the previous section: OLS is accepted because $\hat{\beta}$ is a realization of the random variable $\tilde{\beta}$ whose mean value is identical with β, at least under the restrictions of classical LSD's. This may be formalized one step further.

(D11) We say that y is *the mean value of* $x \in M_m(\text{CLSD})$ ($y = E(x)$) iff x has the form $\langle n, p, d, W, \mathfrak{A}, \mu, \tilde{p}, \tilde{d}, \tilde{u}, \beta\rangle$ and $y = \langle n, p, d, W, \mathfrak{A}, \mu, \tilde{p}, \tilde{d}, \tilde{u}, E(\tilde{\beta})\rangle$

(T3) If $x \in M_m(\text{CLSD})$ then $E(x) \in M(\text{LSD})$

Proof. Trivial by (T1).

Prima facie we could omit completely the theoretical apparatus

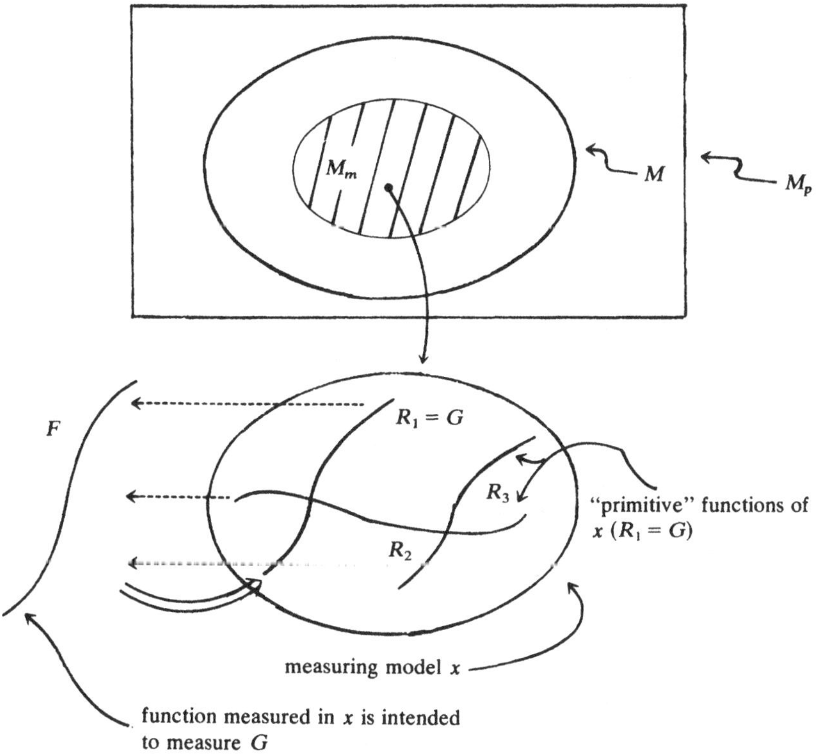

Fig. 3.

together with the probability space in the measuring models which would leave us with structures of the form $\langle n, p, d, \hat{\beta}(p, d)\rangle$. But then the statistical justification for the particular choice of $\hat{\beta}$ could not even be stated for then we lack of the true value β, and the theoretical condition of linearity which is essential for making β "true".

These considerations directly lead us to a central difference between the natural and the social sciences: in the social sciences the models of the respective theories exhibit some distance to observational or experimental results. In order to account for OLS as a method of measurement the criteria for acceptance known from the natural sciences have to be generalized and weakened.

The acceptance and justification of methods like OLS implicitly refers to some counterfactual argument:

(2) If we could collect further data for different, and many events $w \in W$ the mean-values of the corresponding $\tilde{\beta}_i(w)$ in the long run would converge (with probability 1) towards β_i.

In the natural sciences the same argument is *not* counterfactual. There, it simply describes what actually happens. In the natural sciences we are in fact able to repeat experiments with the "same" system (philosophical objections notwithstanding), and in this way to realize the antecedent of (2). This difference between the phenomena dealt with by natural and social sciences is of course well known, but it is only in connection with concepts of measurement that it may be formulated sharply.

It is tempting here to introduce the notation of a repeated experiment given by a sequence $w_1, w_2, \ldots$ in W and corresponding data $\tilde{d}(w_i)$, $\tilde{p}(w_i)$, and to refer to corresponding convergent sequences of mean-values. But this account would introduce an element which, in fact, is not operationally accessible and has no real referent in many situations in the social sciences. Such convergent sequences are typical for natural science but in the social sciences they often simply are not feasible. The only way to obtain such sequences in the social sciences is to increase the length of the time-series (n in (D6)). But social systems change quickly over time, and there is no warrant for regarding such a sequence as a repetition of the "same" situation. In the natural sciences, by contrast, the systems are stable enough to regard a "time series" of observations as "repetitions of observations of the same system". And this is of course the basic justification for applying probabilities and statistics.

V. REGRESSION METHODS OF MEASUREMENT

The kind of measurement studied here may easily be generalized. We do not attempt to give the most general formulation. Sticking to cases in which the determination of real parameters is at stake our definition still covers a wide range of examples in which OLS is applied. Roughly, the generalized measuring models which we call regression measuring models are measuring models in the sense of Section I for some parameters $\hat{\beta}$ which are added to the models of a given theory

((D12–a) below). $\hat{\beta}$ is required to be definable in terms of the other components of the measuring model (D12–a–1). Furthermore, the model should contain a probability space $\langle W, \mathfrak{A}, \mu \rangle$ (D12–a–4), and for some estimators $\tilde{\beta}$, also definable in terms of the components of the measuring model (D12–a–5), the value $\hat{\beta}$ actually measured should be a realization of the estimator (D12–a–5.3). A final, central condition is that the mean value of $\tilde{\beta}$ when replaced for the "true" parameters β in the original model $x[\beta]$ should yield a model of the theory (D12–a–5.4). This condition, which we call model replacement condition, is satisfied, for instance, whenever $E(\tilde{\beta}) = \beta$.

(D12) Let T be a theory with classes M_p and M of potential models and models, respectively.

(a) x is a *regression measuring model* for β *in* T relative to $B, \Sigma, \tau, W, s, \hat{\beta}, D$ iff

(1) x is a measuring model for the defined term $\hat{\beta}$ characterized by $B, \Sigma, \tau, =$ and D,

(2) $s \in \mathbb{N}$, and $\beta, \hat{\beta} \in \mathbb{R}^s$,

(3) x has thc form $x[\beta]$ and $x[\beta] \subset M$,

(4) W is a component of x and among the components of x there are $\mathfrak{A}, \mu$ such that $\langle W, \mathfrak{A}, \mu \rangle$ is a probability space

(5) there is some $\tilde{\beta}: W \rightarrow \mathbf{R}^s$ such that

(5.1) $\tilde{\beta}$ is definable in terms of the components of

(5.2) β is integrable with respect to μ,

(5.3) there is some $w_o \in W$ such that $\tilde{\beta}(w_o) = \hat{\beta}$

(5.4) $x[E(\tilde{\beta})] \in M$.

(b) M_m is a *regression method of measurement for* $\{\beta\}$ iff there exist B, Σ, τ, s and D such that for all $x : x \in M_m$ there are W and $\hat{\beta}$ such that x is a regression measuring model for β in T relative to $B, \Sigma, \tau, W, s, \hat{\beta}$ and D.

(T4) If M_m is a regression method of measurement for β then M_m is a method of measurement for the defined term $\hat{\beta}$

Proof. Obvious.

NOTES

* We are indebted to M. Kuettner and M. Kuokkanen for helpful remarks on an earlier draft.

[1] Compare (Balzer, 1985) and (Balzer, 1988) for more detailed accounts of this view.
[2] A standard reference on fundamental measurement is (Krantz et al., 1971). There, also various kinds of extensive systems are studied.
[3] Compare (Balzer, 1988) for further details.
[4] Cases in which only "parts" of F are uniquely determined can be treated simply by restricting the process of measurement to those arguments of F for which uniqueness obtains. Note the difference between requiring that F be uniquely determined in terms of B and the components of x different from x, and requiring that the function values of F are uniquely determined by F's arguments and by F. The former requirement amounts to

$$\forall F \forall F'(B(D_1, \ldots, R_n, F) \wedge B(D_1, \ldots, R_n, F') \rightarrow F = F')$$

while the latter means

$$\forall b \forall b'(F(a) = b \wedge F(a) = b' \rightarrow b = b').$$

[5] Compare (Balzer–Moulines–Sneed, 1987) for a detailed account of empirical theories.
[6] Note the close analogy to cases of theory-dependent measurement in the natural sciences. There, the basic laws of a theory usually also are not sufficient to guarantee uniqueness of the function to be measured. Further ad hoc assumptions, which can be drawn from a large stock of possibilities, have to be added in order to obtain measuring models. Compare (Balzer, 1985).

REFERENCES

Balzer, W.: 1985, *Theorie und Messung*, Springer, Berlin.

Balzer, W.: 1988, 'The Structuralist View of Measurement: An Extension of Received Measurement Theories', to appear in the *Minnesota Studies in the Philosophy of Science*.

Balzer, W., C. U. Moulines and J. D. Sneed: 1987, *An Architectonic for Science*, Reidel, Dordrecht.

Krantz, D. H., R. D. Luce, P. Suppes and A. Tversky: 1971, *Foundations of Measurement*, Academic Press, New York.

Schneeweiß, H.: 1971, *Ökonometrie*, Physica-Verlag, Würzburg.

Manuscript received 25 January, 1988

Universität München
Ludwigstr 31, D-800 München 22

WERNER DIEDERICH

THE DEVELOPMENT OF MARX'S ECONOMIC THEORY*

ABSTRACT. Marx develops his economic theory in *Capital* in a rather peculiar way. This paper focuses on some of these peculiarities, especially his attempt to base his account of prices and derivative entities (profit, rate of profit, etc.) on the *labour theory of value*. Although he may be said to have failed in this, there is still some kind of 'Marxist' theory of prices possible. This is due to both, the so-called *fundamental theorem* (linking profit and surplus-value) and the possibility, shown by Sraffa and others, to determine prices from the physical parameters of production. By adding on an earlier paper of mine, in which the surplus-value theory has been reconstructed within the structuralistic framework, this paper sketches such a reconstruction for the basic parts of a full-blown Marxist economic theory of capitalistic production.

1.

In a previous paper[1] I construed part of Marx's economic theory as a sequence of four so-called theory elements $T^0, \ldots, T^3$, introducing the concepts of value, money, labour power, and surplus-value, respectively. I did this within the structuralist framework developed by Sneed, Stegmüller, and others.[2] Hence each theory-element T^j was considered as a pair $\langle K^j, I^j \rangle$ consisting of a mathematical structure K^j and a so-called domain of intended applications, I^j, which is identified with a certain sub-structure of K^j. The claim connected with a theory-element $\langle K^j, I^j \rangle$ is that the "core" K^j is, in fact, applicable to I^j, i.e., that it contains a proper extension of I^j.

In the earlier paper I had focused on the *domains* I^j and their connections. In this paper I am going to expand my reconstruction and to dig into some peculiarities of the development of Marx's economic theory which are mainly connected with the *cores* K^j. I shall do this, however, in a less formal way and, incidentally, by relating Marx's ideas to concepts more familiar to contemporary economists. My account will exhibit certain difficulties and, indeed, severe shortcomings of Marx's theory. I am not mainly interested, however, in the question "Marx right or wrong?". The aim of my considerations is more to find out reasons Marx may have had for presenting his theory in the way he did, although he seems to have anticipated at least some

Erkenntnis **30** (1989) 147–164.

of the problems that modern economists so readily are able to point to after far more advanced mathematical and conceptual tools have been developed. The supposed reasons of Marx I want to exhibit are, however, of a more structural sort than those more historical ones which generations of Marxian and Hegelian scholars have proposed.

To put it more concretely, what I am going to do is this: In four more expository sections I shall indicate the scope of Marx's theory insofar it will be reconstructed (Section 2), recall Marx's basic analysis of the production of *surplus-value* (Section 3), contrast this with his notion of *profit* (Section 4), and state the basic problem that Marx's theory of prices has to face (Section 5). In the second, reconstructing and evaluating, part of this paper (Sections 6–10) I first recall the basic role of the law of value on the pre-dynamical stages of Marx's theory (Section 6), then dwell on the conceptual augmentations necessary for the dynamic part (Section 7), reconstruct the Marxian theories of values and prices (Sections 8 and 9), and summarize, and give an assessment of, the results (Section 10).

2.

The scope of Marx's economic theory, as developed in *Capital*, is, of course, not exhausted by dealing with the concepts of *value*, *money*, *labour power*, and *surplus-value* alone, as captured by the theory elements $T^0, \ldots, T^3$, respectively. One main reason for this is that Marx's theory, as reconstructed so far, i.e., up to the explanation of surplus-value, does not yet deal with his account of *profits* and *prices*, as developed in *Capital*, vol. III. I shall therefore formulate, or at least give a sketch of, a fifth theory element, T^4, supposed to cover Marx's theory of so-called *production prices*.

In this part of the theory Marx tries to account for the fact of an approximately uniform *rate of profit*, i.e., the fact that different industries (with different "organic composition" of capital, cf. below) more or less show the same rate of profit (in terms of money), although, in general, connected with different *value* rates of profit. The rate of profit is, of course, a basic concept for the dynamics of capitalism, indeed so much so, that with the theory element T^4 the reconstruction of the main body of Marx's economics will be completed.

3.

Before incorporating Marx's vol. III theory of prices into our reconstruction let me shortly recall the main features of his vol. I theory of *surplus-value*, as construed in T^3.

Conceive of a capitalist investing a certain capital C_0 into the production of a certain good, a, over a certain period of production, say a year (a is regarded as the total output of the respective kind of good during the production process considered). For part of the capital, the "constant capital", C, he or she buys means of production, a^*; a^* usually consists of goods of several kinds (raw materials, tools, etc.),[3] here just lumped together for analytical purposes. For the other part of C_0, V, the "variable capital", the capitalist hires workers, i.e., he or she buys a certain amount of labour power, l, sufficient for that year's production:

$$\begin{array}{ccc} C_0 = & C & + \; V \\ & \downarrow & \downarrow \\ & a^* & l \end{array}$$

The "addition" of l to the means of production, a^*, yields the product $a^* \oplus l$, i.e., the total output, a, the selling of which gives the capitalist back a capital C_1:

$$\begin{array}{lcccll} C_0 = & C & + \; V & < & C_1 & \text{sphere of circulation (prices)} \\ & \downarrow & \downarrow & & \uparrow & \\ & a^* \oplus & l & = & a & \text{sphere of production} \end{array}$$

Usually C_1 is larger than C_0, as indicated in the diagram. (Otherwise, if the capitalist could reasonably have expected $C_0 \geqslant C_1$ he or she would not have taken the risk to invest his or her capital.) Anyhow, the observed fact that capitalists, in general, realize some profit is taken for granted and constitutes, in a rough form, the explanandum.

So far we have not done much more than to *describe* the phenomenon. What goes beyond a mere description is, however, that the buying and selling procedures (the arrows in the diagrams) are understood by Marx as exchanges of *equivalents*, i.e., exchanges of commodities *according to their values*. This cuts off a vulgar sort of "explanation" of the capitalist's profit, namely by tracing it back exclusively to market phenomena and, maybe, a special cleverness, or rudity, on the side of the capitalist. (*Some* capitalists, here and there, may gain some extra-profit from that; but this could not explain the *general* fact of profit making: profit cannot stem, in general, from the sphere of circulation.)

The refined explanandum thus is: how come that the capitalist realizes some profit despite the supposed fact that only equivalents are exchanged? The answer Marx gives is, speaking in terms of the diagram above, that the lower line,

$$a^* \oplus l = a,$$

is no numerical equation at all; the "production operator $\oplus$" (my expression) is no addition (and could not be, because one cannot aggregate "real" goods and labour power). Hence it is not contradictory that the sum of the *values* of a^* and l is smaller than the *value* of the product:

$$v(a^*) + v(l) < v(a).$$

In other words: the value function v is not extensive with respect to $\oplus$.

Note that $v(l)$ is *not* the "value" of the "living" labour that produces a out of a^*, but the value of the respective labour *power*. The labour itself does not appear on the market, hence is no commodity and therefore has no "value" at all! In the eyes of the capitalist the production process may thus appear as an aggregation of values with the wonderful feature that some *surplus-value*,

$$v(a) - [v(a^*) + v(l)] =: S$$

emerges.

4.

Since the capitalist owns the commodities, a^* and l, he or she naturally and legally appropriates the product a, too, which incorporates the surplus-value or, in money terms, the *profit*

$$p(a) - [p(a^*) + p(l)] =: P,$$

where p is the price function.

Now, what the capitalist – and, in order to study capitalism, Marx as well – is mainly interested in is not the absolute profit, but the *rate of profit*

$$r := P/(C + V),$$

i.e., the percentage of profit. Given the mobility of capital, the capitalist might want to choose *that* industry for an investment of his or her money that shows the highest rate of profit. As a matter of fact, however, just this mobility (and competitiveness) of capital leads to (approximately) equal profit rates in all sectors of the respective economy, i.e.,

$$P_i/(C_i + V_i) = P_j/(C_j + V_j)$$

for every two sectors i and j. This fact, however, is not – or only would be under very special circumstances – compatible with a proportionality of values and prices, as may easily be seen by dividing both, nominator and denominator, in the above definition of the rate of profit by V, thus obtaining

$$r = \frac{P/V}{C/V + 1},$$

the so-called *fundamental equation* (in money terms).[4] Here P/V is called the *rate of exploitation* (for obvious reasons, cf. below), and C/V the *organic composition* of capital. Now suppose that prices are proportional to values. In the fundamental equation we then may read P, V and C in either way, as values or amounts of money, because only proportions of these magnitudes matter. Since the organic composition is usually different for different industries, i.e.,

$$C_i/V_i \neq C_j/V_j \qquad \text{for } i \neq j,$$

a uniform rate of profit r could obtain only if the rate of exploitation, P/V (defined in value terms), would vary from sector to sector, too. But this would be contrary to the basic assumption of Marx, that all workers are equally exploited, namely by working equally long and being equally paid: for a certain fraction, P/V, of the working day.[5]

Marx was well aware of this difficulty. He nevertheless kept the assumption of the proportionality of prices and values through the whole of vol. I of *Capital*. Evidently he regarded this assumption as a necessary step within his whole account of capitalism. At least he seems to claim that his explanation of the possibility of surplus-value incidentally explains the possibility of profit; in what sense he may be said to be right in this claim will be argued for later. On the other hand, Marx seems to claim also that the value rate of profit, if not the same as the (money) rate of profit, is at least an approximation of the latter; that he is (or would be) wrong in this, will become clear soon.

5.

In vol. III of *Capital*, Marx develops a theory of *prices of production* which closely relates to the uniformity of the rate of profit. Prices of production are just defined in such a way as to give a uniform rate of profit. (The real 'market prices' may differ from the prices of production.) Marx's theory gives an incorrect account of prices of production, however. He probably was mislead by a per chance correct determination of prices in an *example*;[6] his procedure in that example cannot be generalized. It consists, basically, in a multiplication of the output values by the factor $(1 + r)$, r being the value rate of total profit. It can be shown, however, that this rate usually differs significantly from the rate of profit.[7]

One should not blame Marx too much for this mistake which appears as somewhat "natural", given the non-availability of 20th century linear algebra to him. With this convenient tool it is comparably easy to give a correct determination of the prices of production and point to the flaws in Marx's arguments. What is a more severe blow to Marx's theory, however, is the fact that the correct determination of prices does not proceed *via values*, nor is it *possible* to determine them that way other than *post facto*. In other words: the celebrated theory of value proves irrelevant to one of its main purposes (in Marx's eyes): the determination of prices and thereby the alleged explanation of the rate of profit (i.e., its uniformity and, furthermore, its dynamics: its tendential fall etc.). While it is still *possible* (under certain conditions) to determine the values of commodities from the physical conditions of production, it is also possible to determine the prices from the same source, and it is *not possible* to determine them from the values. Given one of the main objectives of Marx's theory, namely to explain prices, profit, rate of profit, and the latter's dynamics within the labour theory of value as a frame of reference, one cannot but concede that Marx followed a *dead end road* in this respect. In a diagram, essentially borrowed from Steedman,[8] the situation of research is as follows:

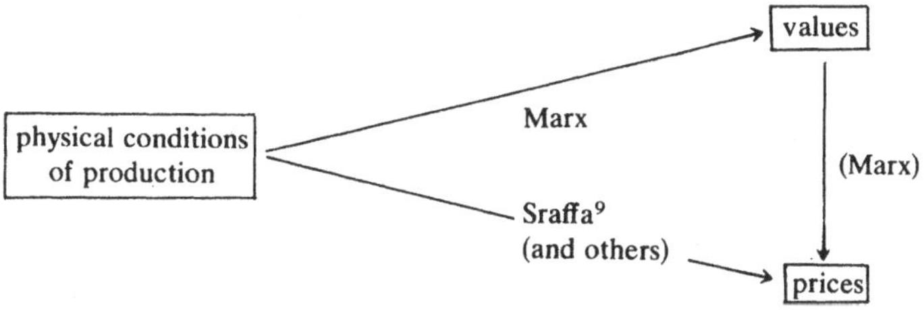

A modern economist may want to comment on this situation: "So what? Prices *can* be determined after all. We just don't need values for that purpose. Why bother? Just forget about values!"[10]

For Marxists, however, there is no such easy way out. Recall that Marx's theory of exploitation crucially hinges on his notion of value: the surplus-value is just the "unpaid part" of the worker's labour; there is no surplus-value and hence no profit without exploitation. But if prices cannot be derived from values, profits cannot be derived from surplus value. Here now modern economic theory comes to help the Marxists with the so-called *fundamental theorem*:[11] The rate of profit is positive iff the rate of exploitation is (i.e., iff there is some surplus-value). This theorem gives at least a global connection of the realms of values and prices – a connection which eventually gives some economic meaning to the category "exploitation" of Marxists' social criticsm. Hence the fundamental theorem appears as somewhat *more* that a consolation price for the overall failure of the genuinely economic parts of Marx's political economy.

6.

Now I turn to the *structuralist reconstruction* of Marx's economic theory (and of a modern emendation of it). I give only some features, however, which bear on the typical kind of development of his theory.

Recall that the fundamental law supposed to connect all steps of Marx's theory is the so-called *law of value*, in our symbolization

$$(LV) \quad z \curvearrowright v \curvearrowright T/p,$$

where z stands (originally) for a function measuring the duration of production processes, v for the value function, and T and p for the relation of exchangeability of goods and the price function, respectively; the arrow stands for 'determines'.[12]

The first step of the theory (core K^0) introduced the concept of *value via* the law LV, the second step (core K^1) the concept of *money as a commodity*, again *via* LV, but with some additional laws specific for the money good, and step 3 (core K^2) the concept of *labour power as a commodity*, i.e., still again *via* LV, and also with some additional laws concerning specific features of this new commodity; these laws especially concern the way of the labour power's "production", namely by *reproduction*, i.e., by consumption of certain goods. (In order to hit Marx's point here, it might be more appropriate to formulate these special laws simply as describing another "industry", which produces the commodity labour power, rather than to introduce the relation R of reproduction as a new theoretical category like in former reconstructions.[13])

7.

So far Marx's theory could be reconstructed more or less within a constant conceptual framework – "more or less" with respect to possible alternatives of reconstruction of the new concepts of money, labour power, and reproduction, either explicitly as new (theoretical or non-theoretical) terms or as concepts for merely singling out the appropriate domains of intended applications, while identifying these domains themselves as subsets of the same set of "partial models".[14]

Marx's theories of surplus-value and profit (T^3 and T^4), however, definitely call for a conceptual augmentation, for two reasons. Firstly, because from this step on a *dynamic* dimension enters the theory, if only in the modest sense that the description of an elementary process of production of surplus-value ("elementary" not with respect to scope, but in the sense of "conceptually elementary", cf. the very beginning of Marx's chapter on commodities) requires the comparison of two temporarily separated states of an economic system. (Nothing is said about the "temporal distance" of the two states; it may be a day or a year or whatever, depending on the circumstances. If, on a later stage of the theory, a dynamical theory in a fuller sense is to evolve, this could possibly be done by regarding infinitesimal time intervals and "summing up" the variations of the economy during these intervals. (Cf. the similar procedures taking place in physics and other empirical sciences as well.) The second reason for an augmentation of the conceptual apparatus is the very fact that, given Marx's specific intents and purposes, it really matters who are the *owners* of the commodities produced (and used for production) and exchanged, as well as consumed (especially by the workers) – if Marx's theory is not only to explain the existence of surplus-value or profit as such, but eventually also the accumulation of surplus-value or profit in the hands of a certain class, the capital owners. But since, in our reconstruction, we will stop at the point of elementary production of surplus-value or profit as such, we might as well forget about the owners of commodities (or – like in earlier steps – take owners in consideration only when singling out the appropriate intended applications).

Thus let us stick to a meager "official description" of the systems our theory is about as consisting of sets of commodities (including labour power, not necessarily money), G, physical conditions of their production, z, and exchangeability relations or prices, T or p. It goes without saying that the commodities are owned by somebody, in particular that products are owned by the owner of the constituents of the product, i.e., the means of production (raw materials, tools) and the labour power involved. After these preliminaries we may say that the kind of minimal description of an economic system to be considered here (for T^3) is a quadruple

$$\langle G, G', z, T/p \rangle$$

instead of the familiar systems

$$\langle G, z, T/p \rangle$$

of the theory elements $T^0 - T^2$. Here G is the set of goods at the beginning of a certain period (of production) and G' the set of goods at the end of that period. (For the time being we do not really need the last component, T or p, because we shall focus here exclusively on the realm of production.) G is transformed into G' partly by consumption and partly by production.

For an illustration let us suppose that there are just three kinds of goods, besides labour power, which are thus transformed. We may think of these three kinds of goods, for instance, as corn (consumption good), iron (means of production), and gold[15] (luxury good to be accumulated or used as money). During the period considered goods of all three kinds are produced, but only iron is used as means of production, i.e., in all three industries the inputs are certain amounts of iron and labour power. Let us further assume that there is as much iron produced as used up in production of all three sectors together, but that there is more corn produced than consumed (by the workers), and that there is a net outcome of gold, too. These specifications (using an illustrative example of Steedman's)[16] are only for giving a concrete idea of the kind of systems considered. More abstractly, we may conceive of a system of production as given by an input-output matrix (including the inputs of labour power) which specifies the bundles of goods before and after the production process, respectively. To be somewhat more precise, let us suppose that there are n kinds of goods. The goods of each kind are measured by a certain unit. Thus the whole set G of commodities at the beginning of the period considered may be regarded as consisting of certain amounts $a_1, \ldots, a_n$ of the various kinds of goods plus an amount l of labour power; $a_1, \ldots, a_n$ are owned by the capitalists, while l is owned by the workers:

owned by capitalists/by workers

$$G: a_1, \ldots, a_n / l$$

Now suppose that there is a fixed technology of production, that the workers get a fixed bundle of commodities as a real wage (or buy a fixed bundle of commodities, if they get their wage in terms of money), that there is no fixed capital, and that there are no joint products.[17] Then we may describe the whole process as follows:

> The labour power, l, is exchanged against a bundle $b_1 \circ \cdots \circ b_n$ ('$\circ$' standing simply for concatenation) of real wage goods (b_j being of kind j, of course), while the rest of the goods a_j, say c_j, are taken as means of production. (We neglect consumption of the capitalists, or regard it as part of their accumulation, to be considered in short due.)

The starting goods of the capitalists are thus split up into a (physical) "constant capital", C^{φ}, plus a (physical) "variable capital", V^{φ}:

$$\begin{array}{llll} g: & = a_1 & \circ \cdots \circ & a_n \\ \| & \quad \| & & \| \\ C^{\varphi} & = c_1 & \circ \cdots \circ & c_n \\ \circ & \quad + & & + \\ V^{\varphi} & = b_1 & \circ \cdots \circ & b_n \end{array} \qquad \text{(exchanged with } l\text{)}$$

(g stands for the capitalists' goods, lumped together.) The parts c_j of the physical constant capital are in turn split up and distributed over the various industries; and so is the labour power acquired:

$$
\left.\begin{array}{lccccc}
 & c_1 \circ \cdots \circ & c_n & \circ & l & =: g^+ \\
 & \| & \| & & \| & \\
\text{industry 1:} & a_{11} \circ \cdots \circ & a_{1n} & \oplus & l_1 & =: a'_1 \\
 & + & + & & + & \circ \\
\vdots & \vdots & \vdots & & \vdots & \vdots \\
 & + & + & & + & \circ \\
\text{industry } n: & a_{n1} \circ \cdots \circ & a_{nn} & \oplus & l_n & =: a'_n
\end{array}\right\} g' \quad \begin{array}{l}\text{(capitalists'}\\ \text{goods after}\\ \text{production)}\end{array}
$$

(g^+ is g after exchange of l with the b_i, but still before production.)

The means of production for industry i: $a_{i1}, \ldots, a_{in}$, are brought together and "combined" with industry i's labour power, l_i, to form the product a'_i. At the same time the workers of all industries, while producing the a'_i's, consume their wage goods b_j, thus reproducing their labour power l (or, to be more precise, thereby "producing" a labour power l' equivalent to l); we may thus regard this "consumptive production" as an $n+1^{st}$ industry, besides the n industries which together "productively consume" the same (amount of) labour power:[18]

$$\text{industry } n+1\text{: } b_1 \circ \cdots \circ b_n \rightarrow l \text{ (or } l'\text{, equivalent to } l\text{).}$$

While l is just reproduced, the n other industries usually produce some *net outputs* (physical surplus goods)

$$s_i^{\varphi} := a'_i - a_i \geqslant 0.$$

At least $s_1 \circ \cdots \circ s_n$ should be more than nothing; otherwise the capitalists would not have invested their goods. Thus we have as output of all $n+1$ "industries"

$$G'\text{: } a'_1, \ldots, a'_n; l' \qquad \text{with } l' \approx l,\ a'_i \geqslant a_i$$

and

$$a'_1 > a_i \text{ for at least some } i.$$

The whole process is given by the inputs and outputs of the $n+1$ "industries", i.e., the matrix (a_{ij}), and the vectors (l_i), (b_i) and (a'_i). (For an account of the technology of the process we may disregard (a'_i) by way of scale-transformation.)

Since these data determine the labour times embodied in each commodity – by solving the appropriate set of simultaneous equations – we may "identify" these data with the function z considered earlier for more simple cases. Thus "z" is to denote the $n \times (n+2)$ matrix of the inputs. In the limiting case, where all goods are directly produced "out of nothing", z could just be given as a vector of the labour times "consumed" in the production of the units of the various goods, and hence as a function $G \rightarrow R^+$ assigning the amount of labour incorporated in every good, as conceived in reconstructing the previous theory-elements; the input/output matrix thus appears as a natural extension of the earlier notion to cover more involved ways of production.

8.

Given the physical data of production, expressed in G, G', and z, we may construe Marx as claiming in T^3 that there is a "value function", $v: G \rightarrow R^+$, obeying the law of value, i.e., being determined by z (and determining in turn T or p), further obeying certain other laws concerning money (if existent) and labour power, namely the laws defining K^1 and K^2, and finally explaining the existence of a net output (over and above the amount of goods consumed by the workers). To be somewhat more precise, T^3 should reproduce in value terms the physical inequalities noted above, i.e., v should fulfill $v(a_i') \circ v(a_i)$ for all i, $v(a_i') > v(a_i)$ for at least some i, and $v(l') = v(l)$, despite the supposed exchange of equivalents only. To put it in a slogan, profit is possible on the basis of freedom and fair exchange.

Thus far what the theory claims of a production process is only that a net outcome of goods is accompanied by a net outcome of values, i.e., not much. The point of Marx's theory of surplus-value, however, is not that there is *somewhere* a surplus-value, but that there is a surplus-value in the hands of the capitalists, *although* there is an exchange of equivalents only, i.e., an exchange according to values (or: at prices proportional to values). In order to bring this point out appropriately in our reconstruction we should perhaps change our mind and enlarge our conceptual apparatus by incorporating the owners of commodities and the actual exchanges between them, especially on the labour market. (*Die Waren können sich nicht selbst zu Markte tragen.*) But again, we may conceive of these circumstances as expressed already in the very description of our system: that at time 0 all means of production and all labour powers are exchanged, i.e., bought by the capitalists, so that at time 1 the whole product belongs to them. Hence the value of the surplus-product, i.e., the surplus-value, belongs to them, too.

In terms of the diagram considered above (Section 3), the production of surplus-value reads:

$$\begin{array}{lcccccl} C_0 = & C & + & V & < & C_1 & \text{sphere of circulation (prices)} \\ & \downarrow & & \downarrow & & \uparrow & \\ & v(a^*) & + & v(l) & < & v(a) & \text{sphere of production} \end{array}$$

Returning to this old scheme, we may say that Marx's theory locates the emergence of the surplus-value clearly on the lower floor, the level of production. Physically there are – in general – commodities of several sorts involved, usually various kinds of means of production, plus labour power. In physical terms, therefore, it is generally not evident what part of the outcome is the surplus product. In order to find that out it is necessary to find a measure which allows to aggregate the various input commodities and compare them with the outcome product. Marx's thesis is that labour value is such a measure and, furthermore, that this measure incidentally regulates the exchange of commodities.

Given this measure v, we can aggregate the means of production and the labour power into value quantities C^v and V^v, respectively. If g'^v is the value quantity of the product, we thus have that the value g^v of total input is smaller than the value of the output:

$$g^v = C^v + V^v < g'^v,$$

i.e., there is a *surplus-value*

$$S := g'^v - (C^v + V^v) > 0.$$

If we assume, like Marx did in *Capital*, vol. I, that prices are proportional to values, we can also read this as saying that there is a positive *profit*:

$$P := C_1 - (C + V) > 0, \qquad \text{Q.E.D.}$$

9.

T^3 is only an intermediary step in Marx's analysis of capitalist economy. What he aims at, and in *Capital* vol. III focuses on, is the dynamics of capitalism and, therefore, a theory of prices, (money) profit, and (money) rate of profit. The foundation of these later parts of his theory are reconstructed in a further theory-element, T^4. As explained earlier, Marx failed in this part of his theory. The correct theory that should replace Marx's will be reconstructed as a theory-element alternative to T^4, $T^{4\prime}$. I shall argue that $T^{4\prime}$ is, in a sense, still "Marxist"; hence it seems possible to picture the main lines of Marx(ist)'s economic theory by the following diagram:

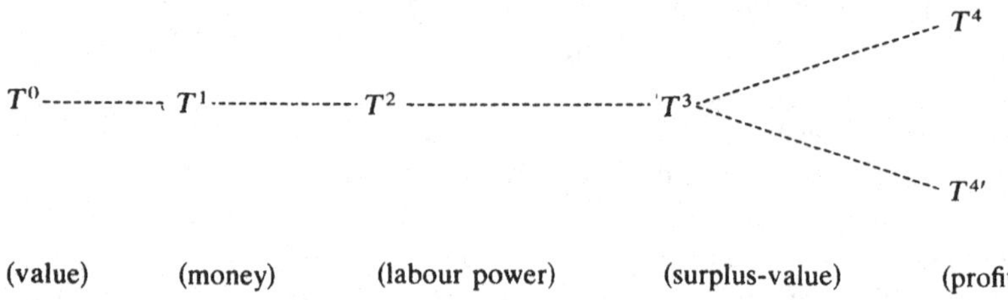

(The linearity of $T^0 - T^3$ is not essential; since wages may be paid in real goods, and even capitalist production of surplus-value is conceivable for economies without money, the diagram could take this form, too:)

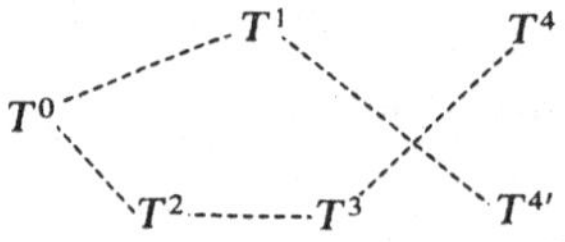

I give only sketches of T^4 and $T^{4\prime}$. Both theory-elements are again about economic systems

$$\langle G, G', z, p\rangle$$

(or systems explicitly described by concepts of owners of commodities as well). These systems are supposed to show a uniform (money) rate of profit

$$r = P/(C + V),$$

where P, C, and V are quantities of money. Marx's claim about such a system is that there exists a value function v, obeying the laws of the previous cores and, furthermore, determining the prices of commodities in such a way that they yield a uniform rate of profit. (If we single out, as indicated, only such economies as intended applications, which show this feature of uniformity, we may just require that the value function determines the prices. This, in fact, seems to be more akin to Marx's intentions, because his account of prices should be wide enough to cover also other phenomena like the dependence of prices on supply and demand, degree of competition, mobility of capital and labour power, and so on. These issues, however, for which a full reconstruction of Marx's theory should conceive of further theory-elements T^5, $T^6, \ldots$, is out of the scope of the present discussion. In other words, we are considering here only so-called *prices of production*, i.e., ones that show a uniform rate of profit, no market prices.[19]

Marx's attempted solution to the problem supposedly solved by T^4 consists basically in multiplying the values of outputs by the factor $(1+r)$, where r is the total value rate of profit, $S/(C^v + V^v)$. But, besides Marx's failure to transform the values of the inputs, too, his procedure cannot work (or only, per chance, in some suitable cases), because the value rate of profit is, in general, different from the (money) rate of profit, as can be shown by a correct analysis of prices.[20] This correct analysis, giving the basis for $T^{4'}$, consists simply in augmenting the input/output scheme of the economy, including worker's consumption, by the respective price factors. This leads to a number of equations, say $n+2$, for the (money) rate of profit, r, the (money) wage (per unit of labour power), w, and n prices of units of goods other than money (the price of money being 1 by convention). Solving these equations gives r, w, and the prices of non-money goods on the basis of the physical data of production. Incorporating the "fundamental theorem" into $T^{4'}$ leads to the claim: there is a value function v such that v obeys the law of value and the other laws of $T^0 - T^3$, and the physical data z are connected with the prices p in such a way that there is a positive profit iff there is a positive surplus-value. (We may simplify this formulation of $T^{4'}$'s laws by the understanding that the determination of p by z is a shortcut of LV suggested by the latter's interpretation as a functional equation $p = \Psi(\Phi(z, \ldots), \ldots)$, namely a shortcut $p = \Xi(z, \ldots; \ldots)$.[21] Not incorporating the fundamental law into $T^{4'}$ would result in a core $K^{4'}$ without theoretical components, contrary to the dependence of $T^{4'}$ on T^3 suggested by the diagram above. More substantially: $T^{4'}$ with just $p = \Xi(z, \ldots)$ would be isolated, not expressing that the kind of economies that $T^{4'}$ is to cover presupposes the "subsumption" of (money and) labour power under the LV, i.e., the market of commodities.)

We have thus finished the reconstruction of a skeleton of Marx's economic theory, as developed by him in *Capital*, vols. I and III, and partly corrected by more recent economists. We have seen that Marx's original theory has some severe flaws which suggest a sharp bend of the theoretical development towards an elimination of value terms. In a more general sense, however, Marx's materialist intentions are kept in that prices, profits, and so on are shown to depend on quantities rooted in the physical conditions of production; profits are thus shown not to stem from the sphere of circulation where alone they "appear". In this sense, and in Marxian terms, one thus still might say that profits, the driving force of capitalistic economy, are based on surplus-labour and hence on exploitation.

10.

Marx's theory was evidently supposed to follow some inner logic. The basic logical tie of the various stages of the theory is the *law of value*. This law is supposed to rule the hidden connections between two realms or "spheres": that of prices (or "circulation") and that of production.

One of the main purposes of Marx's, if not *the* main purpose was to explain the dynamics ("Entwicklungsgesetze") of capitalism. These crucially hinge on the rate of profit which rules the decisions of the capitalist. Despite this main focus of Marx's he found it necessary first to develop a theory of surplus-value and exploitation in order to explain the possibility of profit as such. He did this in vol. I, as reconstructed in $T^0, \ldots, T^3$.

Volume III was to continue this line of thought to the genuine subject of his studies: a theory of prices and dynamic features of capitalism depending thereon. Marx failed, however, in this part of his theory, T^4. Later developments of economic theory show: a theory of prices is possible, though in a different way, i.e., by $T^{4'}$, taking a shortcut from production to prices, so to speak. Nevertheless there are connections with the theory of value, which allow to say that while it is not necessary (and not even possible) to *calculate* prices from values, the *sources* of profit lie in surplus-value and hence in *exploitation*.

A final remark on the *rationale* of a *structuralist* reconstruction, as the one given, may be helpful. Some achievements of this approach to Marx's economic theory have been listed already in the earlier paper mentioned at the beginning.[22] Thus let me only emphasize again that the kind of continuity of theoretical development germane to the connecting *law of value* may not likely be depicted within the statement view of theories. Note that the concepts involved in this law do change their meanings while the theory progresses. Also the connections claimed by the law get more and more specific or complex (depending on which alternative of reconstruction one chooses[23]). In this paper this line of thought has been continued with a broadening of understanding of z which now stands for the physical conditions of production in general.

The "empirical" line of development, i.e., the development of the domains of intended applications, has not been focused on in this paper. What has been said about that in the earlier paper may be continued, too, with respect to the new theory-elements, T^4 and $T^{4'}$. The respective domains of application are very different, namely those economic systems which show a uniform organic composition of capital and those systems which are not restricted in this way (but show other idealized properties assumed, but not discussed, throughout the whole reconstruction), respectively; i.e., the domain of applications that reasonably may be envisaged for T^4 is very small, if not empty, while that for $T^{4'}$ is the one that Marx actually has intended for T^4, but, as we now know, cannot be upheld for T^4.

Let me finally stress again the point, that the structuralist account of Marx's economic theory avoids any hypostasization of the concept of value as found in Marx himself. In our reconstruction *value* appears only as a component v of certain models. These models themselves never serve as subjects of the theory. What the theory is *about* (in our reconstruction) are certain systems described by empirical quantities alone. The labour theory of value claims nothing more (nor less) than that certain features of economic systems, which may be described entirely without recourse to the concept of value, are such that they can be regarded as parts of richer structures including the value component. No talk of "substance", "essence", and the like is necessary for that purpose.

NOTES

* I am indebted to Prof. Peter Flaschel for some useful hints.

1 Diederich[82].

[2] Cf. the references in the paper mentioned above (Note 1). In this paper I try to get along with even less logical technicalities.
[3] For the term 'means of production' cf. Cohen [78], p. 38.
[4] Cf. Elster [85], 3.2.2, p. 133.
[5] Cf. Elster, *ibid.*; cf. however recent discussions by U. Krause and R. Picard, see P. Flaschel [83].
[6] Cf. Elster, *op. cit.*, p. 134.
[7] Cf. e.g., Steedman [77], Chap. 3, p. 46.
[8] *Op. cit.*, p. 48.
[9] Sraffa was probably the first dealing in full generality with the derivation of prices from the physical conditions of production.
[10] Cf. Elster's diagrams, showing the dismissability of values, *op. cit.*, p. 138.
[11] Morishima and others, cf. Elster, *op. cit.*, 3.2.3, p. 141.
[12] Cf. the paper mentioned Note 1 above.
[13] Cf. Section 7 below and the paper mentioned Note 1 above.
[14] I.e., $M^0_{pp} = M^1_{pp} = M^2_{pp}$, cf. ref. Note 1.
[15] Or – to make the illustration less dull – theology books (after a proposal of R. P. Wolff's, cf. Wolff [84], p. 14).
[16] Steedman, *op. cit.*, Chap. 3.
[17] Cf. Steedman, *op. cit.*, e.g., 202.
[18] Cf. Section 6 above and Wolff, *op. cit.*, Chap. 1.
[19] Cf. Steedman, *op. cit.*, p. 45.
[20] Cf. Steedman, *op. cit.*, pp. 43–45.
[21] Cf. Diederich [81], 5.1.4, p. 133f.
[22] See ref. Note 1, p. 157.
[23] See again ref. Note 1.

REFERENCES

Cohen, G. A.: 1978, *Karl Marx's Theory of History*, Clarendon, Oxford.

Diederich, W.: 1981, *Strukturalistische Rekonstruktionen*, Vieweg, Braunschweig/Wiesbaden.

Diederich, W.: 1982, 'A Structuralist Reconstruction of Marx's Economics', in W. Stegmüller, W. Balzer and W. Spohn (eds.), *Philosophy of Economics*, Springer, Berlin, pp. 145–60.

Elster, J.: 1985, *Making Sense of Marx*, Cambridge University Press, Cambridge.

Flaschel, P.: 1983, *Marx, Sraffa und Leontief*, P. Lang, Frankfurt/M

Steedman, I.: 1977, *Marx after Sraffa*, NLB, London.

Wolff, R. P.: 1984, *Understanding Marx*, Princeton University Press, Princeton.

Manuscript received 25 January, 1988

Abteilung Philosophie
Universität Bielefeld
Postfach 8640
D-4800 Bielefeld 1
West Germany

MAARTEN C. W. JANSSEN*

STRUCTURALIST RECONSTRUCTIONS OF CLASSICAL AND KEYNESIAN MACROECONOMICS

In recent years a lot of authors have shown considerable interest in reconstructing economic theories, especially microeconomic general equilibrium theory, in a structuralist way. Up to now almost no effort has been undertaken to analyse macroeconomic theories along structuralist lines. It is the aim of this paper to fill this gap by presenting structuralist reconstructions of three macroeconomic theories. Special attention will be given to the way *economists* use these theories in order to interpret national economic statistics. By comparing the reconstructions some surprising similarities, especially with respect to the equilibrium character of the three theories, will become apparent. It will turn out that the only difference in structure between classical and Keynesian theory is with respect to the labour market.

In order to be able to indicate the similarities in structure between the theories it is necessary to devote a first section to some definitions of the notion of equilibrium in economics. Section 2 deals with a simple macroeconomic textbook theory and its reconstruction. Though the simple theory is only a starting-point for more comprehensive theories it contains already the basic structure of the classical and Keynesian theories, which together with their reconstructions will be discussed in Sections 3 and 4. The fifth section contains some general remarks concerning the reconstructions presented in the earlier sections and the way they are related to reconstructions of general equilibrium theory. The last section treats the question whether there are theoretical terms in macroeconomics. I will show that two different positions with respect to this question are defensible and that both positions are compatible with the way economists relate the concepts that enter their theories to national economic statistics. Though the differences between the structures of classical and Keynesian theories turn out to be small they are important enough to create differences of opinion with respect to the question how to interpret labour market statistics.

Erkenntnis **30** (1989) 165–181.

1. THREE DEFINITIONS OF THE NOTION OF EQUILIBRIUM

Economists have a habit to use the same word for different conceptions. This is also the case with the notion of equilibrium (Machlup, 1963). The equilibrium concept is frequently associated with the equilibrium of general equilibrium theory in which supply and demand balance on all markets. In this vein Keynesian theory is often viewed as a disequilibrium theory because it assumes that the labour market is not cleared. Others have argued that Keynesian theory is an equilibrium theory too, but that another notion of equilibrium is involved. As the notion of an equilibrium plays a crucial role in the rest of the paper I will distinguish several ways in which the concept of an equilibrium can be used. In the following three sections I will discuss in what sense the respective theory is a (dis)equilibrium theory.

The three[1] concepts of a state of equilibrium that play a role in the sequel are the following:

(i) A certain state of the economy is an equilibrium state if the desired actions of (groups of) individuals are consistent with each other (Bénassy, 1982). An action is desired if it is the most preferred action *subject to the ruling restrictions the (group of) individual face(s)*. Desired actions can be carried out if they are consistent with each other.

(ii) A certain state of the economy is an equilibrium state if demand equals supply on all markets. This is the equilibrium notion associated with general equilibrium theory. It is a specialization of the first definition in the sense that the only restriction the (group of) individuals face(s) is the budget constraint.

(iii) A certain state of a system is an equilibrium state if "the relations that describe the economic system form a system sufficiently complete to determine the values of its variables" (Arrow, 1968).

One has to note that there are no dynamic considerations in the above equilibrium notions. *If* time plays a role in a theory one could define an equilibrium as a situation of rest, i.e., as a situation in which there is no change in time. Such an equilibrium is usually called a stationary equilibrium (Fisher, 1936). As there is no time involved in the theories considered in this paper the equilibria are not necessarily stationary equilibria.

2. A SIMPLE MACROECONOMIC THEORY

In order to be clear about the exact nature of the theories I will reconstruct this and the next two sections start with a short presentation of the respective theories in the economists' way.

Simple macroeconomic textbook theories, as the one presented in this section, study the interaction between three different sectors (households, firms and government) in two markets: the product market and the money market. The simple theory resembles to a large extent Hicks' 1937-interpretation of Keynes' General Theory and is, since then, generally accepted as a way to obtain Keynesian (unemployment) results.

All three sectors participate in the *product market of the economy*. Households consume, save and pay taxes; firms produce and invest; the government receives taxes and purchases some goods. It is assumed that consumption demand is a function of disposable national income, taxes are functionally related to national income, business' desire to invest is a function of the real interest rate, while the demand for goods by the government is exogenous. All variables are measured in real terms. The supply of output is determined by a short-run production function in which the capital stock is fixed. Equilibrium on the product market is defined by the equality of the supply of real output and the respective demands. To sum up,

Demand:	Households' real consumption demand	(1) $c = f_1(y - t(y)) = f_1'(y)$
	Firms' real investment demand	(2) $i = f_2(r)$
	Government's demand	(3) $g = g_0$
Supply:	Firms' real output	(4) $y = f_4(k_0, n) = f_4'(n)$
Equilibrium:	Demand equals supply	(5) $c + i + g = y$

where c is real consumption demand, y is real national income/real output, t is taxes, i is real investment demand, r is real interest rate, g is real government demand, k is capital stock, n is the level of employment.

In the *money market of the economy* a demand for and a supply of money balances can be distinguished. The demand for real money balances depends on the real interest rate (the speculative demand) and national income (the transaction demand). In an elementary theory, as I consider here, the nominal supply of money is exogenous. Equilibrium on the money market is defined by the equality of the demand for and the supply of real money balances. To sum up,

Demand: Real speculative and transaction demand for money (6) $m^d = f_5(r, y)$
Supply: Nominal supply (7) $m^s = m_0^s$
Equilibrium: Supply equals demand in real terms (8) $m^s/p = m^d$

where m^d is the real demand for money, m^s is the nominal supply of money, p is the price level.

It is easy to see that all the demand and supply variables can be expressed as functions of the real interest rate and the level of employment by using Equation 4. The two equilibrium conditions contain the three variables r, p and n. Hence, the system is indeterminate and there can not be an equilibrium in the third sense of the first section. However, if the price level is held constant (or any other of the three variables) there are only two unknowns and the system is determinate. In this case one can depict the usual IS and LM-curves in (r, n)-space. The IS-(LM)-curve depicts those combinations of r and n at which the product (money) market is cleared.

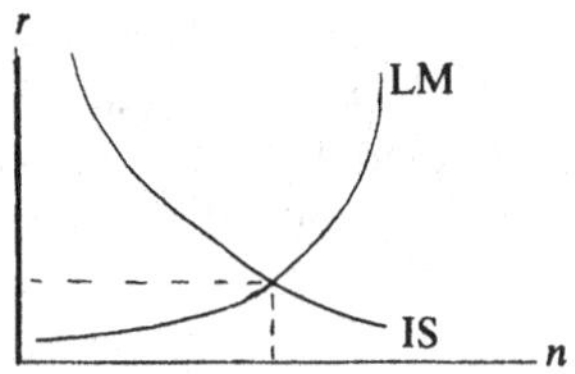

Fig. 1. Equilibrium on the product market and the money market.

Now, I come up with a *structuralist reconstruction*. We have seen that the situation on the two markets[2] is completely determined by the values of r and n *if p is held constant*. The values of r and n determine, so to speak, the state of the system and I will call them 'state values'. In order to make this idea more precise I interpret r and n as single numbers representing one single state of the system for a given p. I will assume that the state values are positive and I introduce a 'state space', i.e., the space of all possible state values, which is simply the Cartesian product of the positive real numbers. Now, it will be clear that all demand and supply variables can be expressed as functions from the state space to the nonnegative real numbers. As all the elements of the potential models of the theory are discussed I characterize the potential models in a formal way.

DEFINITION 2–1. x is a potential model of the simple macro-theory ($x \in M_p$(SMT)) if there are M, r, n, c, i, g, y, m^d, m^s, p such that

(1)	$x = \langle M, r, n, c, i, g, y, m^d, m^s, p\rangle$	
(2)	$M = \{m_1, m_2\}$	set of markets
(3)	$z = (r, n) \in (\mathbf{R}^+)^2$	state values
(4)	$c: (\mathbf{R}^+)^2 \to \mathbf{R}^+$	real consumption demand
	$i: (\mathbf{R}^+)^2 \to \mathbf{R}^+$	real investment demand
	$g: (\mathbf{R}^+)^2 \to \mathbf{R}^+$	real government expenditure (usually a constant function)
	$y: (\mathbf{R}^+)^2 \to \mathbf{R}^+$	real output/production
	$m^d: (\mathbf{R}^+)^2 \to \mathbf{R}^+$	real demand for money
	$m^s: (\mathbf{R}^+)^2 \to \mathbf{R}^+$	nominal supply of money (usually a constant function)
(5)	$p \in \mathbf{R}^+$	price level

Having defined the potential models of the theory there is a rather natural way to proceed to the models of the theory by imposing that the state values have to be such that the two equilibrium conditions are fulfilled.

DEFINITION 2–2. x is a model of the simple macro-theory ($x \in M$(SMT)) if there are M, r, n, c, i, g, y, m^d, m^s, p such that

(1)	$x = \langle M, r, n, c, i, g, y, m^d, m^s, p\rangle$	
(2)	x is a potential model of the simple macro-theory	
(3)	$c(z) + i(z) + g(z) = y(z)$	equilibrium on the product market
(4)	$m^d(z) = m^s(z)/p$	equilibrium on the money market

If the price level is held constant in the simple macroeconomic theory it is easy to see that the equilibrium state of the system is an equilibrium in all the three senses distinguished in the first section: the system is determinate, the equilibrium conditions are of the form 'demand equals supply' and, hence, there is a consistency of desired actions.

3. A CLASSICAL EXTENSION

Both extensions of the simple theory I will discuss treat the labour market. In this section I will discuss the classical treatment of the labour market; the next section deals with the Keynesian treatment. The classical macro-theory can be seen as an application of general equilibrium theory to the context of macroeconomics: in equilibrium all markets have to be cleared. The Keynesian (unemployment) results of the simple theory are, from a classical point of view, due to the fact that a treatment of the labour market is missing.

The classical view of the labour market treats demand for, and supply of, labour in an explicit way. It is assumed, see e.g., Sargent (1979, Chap. 1), that the demand for labour by firms and the supply of labour by households are determined by the real wage rate. Equilibrium on the labour market is characterized by the equality between the demand for, and the supply of, labour. I will present the relations in inverse form in order to be able to get a reconstruction in line with the one presented in the previous section.[3] The economic content of the classical view on the labour market is not affected by this 'inverse presentation'. The 'inverse' of the demand for labour is the real wage firms are willing to offer in order to attract n units of labour. The 'inverse' of the supply of labour is the real wage households demand in order to deliver n units of labour. In this inverse presentation equilibrium is characterized by the equality between the wages households demand and the wages firms are willing to offer.

Demand: Firms' demand for labour	(9)	$w^s/p = f_6(n)$
Supply: Households' supply of labour	(10)	$w^d/p = f_7(n)$
Equilibrium: Demand equals supply	(11)	$w^s/p = w^d/p$

The classical macroeconomic theory is obtained by adding the classical treatment of the labour market to thc simple theory. Thus, the classical theory treats three markets with three equilibrium conditions. It is no longer necessary to treat the price level as exogenous in order to obtain a determinate system. In view of the three equilibrium conditions the price level can be incorporated as a third state value and the state space has to be enlarged to three dimensions. Thus, the models can be characterized as follows.

DEFINITION 3-1. x is a model of the classical macro-theory ($x \in M(\text{CMT})$) if there are $M, r, n, p, c, i, g, y, m^d, m^s, w^s, w^d$ such that

(1)	$x = \langle M, r, n, p, c, i, g, y, m^d, m^s, w^s, w^d \rangle$	
(2)	$M = \{m_1, m_2, m_3\}$	set of markets
(3)	$z = (r, n, p) \in (\mathbb{R}^+)^3$	state values

(4)	$c: (\mathbb{R}^+)^3 \to \mathbb{R}^+$	real consumption demand
	$i: (\mathbb{R}^+)^3 \to \mathbb{R}^+$	real investment demand
	$g: (\mathbb{R}^+)^3 \to \mathbb{R}^+$	real government expenditure (usually a constant function)
	$y: (\mathbb{R}^+)^3 \to \mathbb{R}^+$	real output/production
	$m^d: (\mathbb{R}^+)^3 \to \mathbb{R}^+$	real demand for money
	$m^s: (\mathbb{R}^+)^3 \to \mathbb{R}^+$	nominal supply of money (usually a constant function)
	$w^s: (\mathbb{R}^+)^3 \to \mathbb{R}^+$	nominal wage supply
	$w^d: (\mathbb{R}^+)^3 \to \mathbb{R}^+$	nominal wage demand
(5)	$c(z) + i(z) + g(z) = y(z)$	equilibrium on the product market
(6)	$m^d(z) = m^s(z)/p$	equilibrium on the money market
(7)	$w^s(z) = w^d(z)$	equilibrium on the labour market

(The first four axioms constitute the potential model of the classical macro-theory).

One of the most important features of classical economic theory is the view that real and money markets can be considered separately, i.e., real markets determine relative equilibrium prices and equilibrium quantities of the product and the labour market, while the money market determines only the equilibrium general price level. This view is generally known as the classical dichotomy. In the macroeconomic theory I have presented above this view comes down to the statement that equilibrium conditions of the product and the labour market determine the subset of real state values (the real interest rate and the level of employment), while the equilibrium condition of the money market determines only the price level. Hence, it is possible to make a partition of the state values into two real state values and one monetary state value. Thus, in the case of the classical dichotomy the demand and supply functions of the product and the labour market can be expressed as mappings from $(\mathbb{R}^+)^2$ to $\mathbb{R}^+$, while the demand for and the supply of money can be viewed as mappings from $\mathbb{R}^+$ to $\mathbb{R}^+$. The model of the classical dichotomy macro-theory can be defined by relating the 'real state values' only to the equilibrium conditions of the real markets and the monetary state value only to the money market. By now, it will be clear that the classical dichotomy can be presented as a specialization of the more general classical theory as presented above.

DEFINITION 3–2. x is a model of the classical macro-theory revealing the classical dichotomy ($x \in M$(CDMT)) if there are M, r, n, p, c, i, g, y, m^d, m^s, w^s, w^d such that

(1)	$x = \langle M, r, n, p, c, i, g, y, m^d, m^s, w^s, w^d \rangle$	
(2)	$M = \{m_1, m_2, m_3\}$	set of markets
(3)	$z_1 = (r, n) \in (\mathbf{R}^+)^2$	real state values
	$z_2 = (p) \in \mathbf{R}^+$	monetary state value
(4)	$c: (\mathbf{R}^+)^2 \to \mathbf{R}^+$	real consumption demand
	$i: (\mathbf{R}^+)^2 \to \mathbf{R}^+$	real investment demand
	$g: (\mathbf{R}^+)^2 \to \mathbf{R}^+$	real government expenditure (usually a constant function)
	$y: (\mathbf{R}^+)^2 \to \mathbf{R}^+$	real output/production
	$m^d: \mathbf{R}^+ \to \mathbf{R}^+$	real demand for money
	$m^s: \mathbf{R}^+ \to \mathbf{R}^+$	nominal supply of money (usually a constant function)
	$w^s: (\mathbf{R}^+)^2 \to \mathbf{R}^+$	nominal wage supply
	$w^d: (\mathbf{R}^+)^2 \to \mathbf{R}^+$	nominal wage demand
(5)	$c(z_1) + i(z_1) + g(z_1) = y(z_1)$	equilibrium on the product market
(6)	$m^d(z_2) = m^s(z_2)/p$	equilibrium on the money market
(7)	$w^s(z_1) = w^d(z_1)$	equilibrium on the labour market

(The potential model of the classical macro-theory revealing the classical dichotomy is, of course, characterized by the axioms (1)–(4)). It will be clear that the equilibrium state of the classical macro-theory as well as the equilibrium state of the classical dichotomy is an equilibrium in all the senses distinguished in the first section.

4. A KEYNESIAN EXTENSION

It is well known that Keynesian theory takes another view on the labour market than the classical theory. According to Keynesian theory the nominal wage rate is to a large extent determined independently of economic factors such that it has to be viewed as an exogenous variable with respect to economic theory. It is assumed, see e.g., Sargent (1979, Chap. 2), that the demand for labour is less than the supply of labour at this exogenous nominal wage rate. In this vein the level of employment is determined by the firms' demand for labour, which is related to the real wage rate. As the nominal wage rate is determined by noneconomic forces it can not adjust to the inequality between the demand for, and the supply of, labour. In my 'inverse' notation,

wages are exogenous	(12)	$w^s = w_0$
equilibrium (demand for labour)	(13)	$w^s/p = f_s(n)$

The Keynesian macro-theory is obtained by adding the Keynesian treatment of the labour market to the simple theory. Households' demand for wages (supply of labour) does not play an explicit role in Keynesian theory and it is not included in the list of supply and demand variables in the definition of models of the Keynesian theory. The other differences with respect to Definition 3–1 can be found in Axiom 5 (the Keynesian treatment of wage offers by firms) and Axiom 8 (the form of the equilibrium condition of the labour market).

DEFINITION 4–1. x is a model of the Keynesian macro-theory ($x \in M$(KMT)) if there are $M, r, n, p, c, i, g, y, m^d, m^s, w^s$ such that

(1)	$x = \langle M, r, n, p, c, i, g, y, m^d, m^s, w^s \rangle$	
(2)	$M = \{m_1, m_2, m_3\}$	set of markets
(3)	$z = (r, n, p) \in (\mathbf{R}^+)^3$	state values
(4)	$c: (\mathbf{R}^+) \to \mathbf{R}^+$	real consumption demand
	$i: (\mathbf{R}^+)^3 \to \mathbf{R}^+$	real investment demand
	$g: (\mathbf{R}^+)^3 \to \mathbf{R}^+$	real government expenditure (usually a constant function)
	$y: (\mathbf{R}^+)^3 \to \mathbf{R}^+$	real output/production
	$m^d: (\mathbf{R}^+)^3 \to \mathbf{R}^+$	real demand for money
	$m^s: (\mathbf{R}^+)^3 \to \mathbf{R}^+$	nominal supply of money (usually a constant function)
(5)	$w^s \in \mathbf{R}^+$	nominal wages are a given number
(6)	$c(z) + i(z) + g(z) = y(z)$	equilibrium on the product market
(7)	$m^d(z) = m^s(z)/p$	equilibrium on the money market

(8) $w^s(z)/p = f_s(z)$ equilibrium on the labour market

(Naturally, Axioms (1)–(5) define the potential models of the Keynesian macro-theory).

In the case of the Keynesian theory no specialization exists in which a subset of state values fulfils the equilibrium conditions of some markets. All state values are determined together by the three equilibrium conditions.

It is clear that Keynesian theory is not an equilibrium theory in the sense that demand and supply balance on all markets, i.e., the second sense of the first section. However, it is indeed an equilibrium theory in the two other senses. In particular, the Keynesian system determines the values of all its variables and there is a consistency of desired actions with respect to the exogeneity of nominal wages.

5. GENERAL REMARKS CONCERNING THE PROPOSED RECONSTRUCTIONS

In this section I will make some general remarks concerning the reconstructions I have proposed in the previous sections. The remarks will be concerned with the generality of the reconstructions and the relations to reconstructions of general equilibrium theory. The remarks I will make apply equally well to the simple, the classical and the Keynesian macro-theory.

The reader may be surprised by the fact that the demand and supply variables are treated as functions of *all* state values instead of relating them to specific ones as in the economic presentation at the beginning of each section and as most textbooks present the demand and supply functions. There are two reasons for formulating the reconstructions in a general manner. (1) In the way I have reconstructed the theories the basic structure becomes much more apparent than it would become if a more detailed structure was chosen: for the *structure* of the theory it does not matter whether for example consumption demand is a function of the level of employment only or of the level of employment and the real interest rate. Indeed, Hicks' famous 1937-article ends by proposing a "really general theory" by, in our terms, relating all demand and supply functions to all state values (Hicks, 1937, p. 156–9). (2) Alternative versions of specific demand and/or supply functions can be handled within the same models and potential models. More specifically I will show that for some paradigmatic examples of alternative specifications only new state values have to be included in the economic presentations at the beginning of the previous sections. (i) If real wealth is assumed to effect the demand for consumption purposes economists view nominal wealth as an exogenous value (as the capital stock in Equation 4). Real wealth varies then inversely with the price level and the latter (a state value!) influences consumption demand; see e.g., Branson (1979, p. 206–13). (ii) Sometimes textbooks introduce national income as a second explanatory variable for the demand for investment. Using the short-run production function the demand for investment can be viewed to depend on the level of employment, which is a state value; see e.g., Branson (1979, p. 236–40). (iii) In a more elaborate theory the nominal supply of money is an endogenous variable depending on the real interest rate, which is (again) a state value; see e.g., Branson (1979, p. 271–80).

The second subject of this section concerns the differences and similarities to reconstructions of microeconomic general equilibrium theory by Hands (1985) and Janssen and Kuipers (1988). The similarities are easy to mention. (a) All reconstructions contain a set of markets; in macroeconomics there is a definite set of markets (two or three). (b) On each market there is a demand and a supply function (in microeconomics combined in a market excess demand function); in macro-economics definite demand and supply functions, e.g., consumption demand, demand for money etc. (c) The prices in microeconomics correspond to the state values in the reconstruction of macroeconomics and the associated price space of microeconomics is the state space of macroeconomics. (d) The models of the reconstructions specify the equilibrium conditions.[4]

Besides these similarities there are also differences. Macroeconomics lacks a kind of structure as general equilibrium theory in which there is an individualistic foundation for the market excess demand functions. This is due to the fact that the state values of macroeconomics include, besides 'prices', the level of employment. As not all individuals are employed to the same degree aggregation problems arise if one wants to connect individual demand functions (in which individual employment is an argument) with aggregate demand functions (in which the level of *total* employment is an argument).

Aggregation problems do not arise in general equilibrium theory, because it is assumed that individuals respond to prices only and that prices are the same for all individuals.

The second difference rests on different interpretations of the equilibrium conditions. In general equilibrium theory assumptions are known under which one can prove the existence of an equilibrium price vector. If the assumptions are fulfilled an equilibrium price vector simply exists. Under these assumptions the equilibrium axiom is not a constraint on actual prices: the equilibrium axiom merely states that there is an equilibrium price vector, it does not pose that actual prices are equilibrium prices. The equilibrium axioms of the macro-theories on the other hand do *not only* pose that equilibrium state values exist (they take that for granted), but the equilibrium conditions of macroeconomics have also to be understood as constraints on the possible state values in order to make sense of comparative static analysis.

6. (IN)VOLUNTARY UNEMPLOYMENT: AN EMPIRICAL ISSUE?

After the reconstructions in terms of models and potential models the question arises whether the demand and supply variables are theoretical terms in the structuralistic sense or not. From the controversy between Balzer (1982) and Haslinger (1983) one knows that this question is difficult to handle in economics. In first instance this difficulty is due to the fact that economic theory is essentially ‘out of time’, while applications have to be ‘in time’. In order to avoid a too negative attitude with respect to this issue I will bypass this problem in this paper by doing as if the macro-theories are specified for a definite time period (a year or a quarter of a year or so).

Thus, the question is whether demand and supply can be measured independently of the theory or not. At this point it may be helpful to quote some economists. Weintraub (1985, p. 50) poses that “plans are only coherent if equilibrium is established. We cannot measure incoherent plans” and in Fomby et al. (1984, p. 568) one can read that “an immediate problem is that D_t and S_t (respectively demand and supply in period t–M.J.) are not observable unless $D_t = S_t$”. Thus, these authors may be interpreted as saying that demand and supply are theoretical terms in the structuralist sense, because demand and supply are only observable by using the equilibrium condition of the theory under consideration. However, in the dispute on voluntary versus involuntary unemployment there are economists, especially those who argue in favour for the existence of involuntary unemployment, who do not agree with this position. Solow (1980, p. 3) writes "I believe that what looks like involuntary unemployment is involuntary unemployment" and Standing (1981, p. 576) proclaims that "in economic research there should be a basic presumption that if someone says he wants employment he should be regarded as involuntary unemployed". In my view these authors may be interpreted as arguing that demand and supply are observable independent of the theory under consideration. In this view demand and supply can be revealed by questionnaires. However, these authors have to admit that this method of revealing demand and supply is not a generally accepted practice in applied economics. In this respect the above dispute has not contributed to any noteworthy achievement in applied economics. So, I will leave the dispute for what it is and concentrate in the rest of the section on the question how economists actually measure demand and supply variables.

The economist's parlance to state the problem is the following. Economic theory has an ex-ante character, i.e., it deals with *desired* actions, while the observations have an ex-post character, i.e., they are about *realized* actions. I claim that economists adhere to certain principles which establish relations between the ex-ante and the ex-post values. The first principle is the principle of 'voluntary exchange': one cannot enforce agents to engage in trading goods they do not want. The second economic principle asserts that markets are perfect in the sense that the realized transaction on a market is equal to the minimum of demand and supply on the market, i.e., there can not be unsatisfied demand and supply on the same market. (The 'perfect market'-principle is sometimes called the 'short-side rule'). The third principle says that the theory should apply to the situation at hand.

On *the product and the money market* the principles amount to the following relations between ex-ante and ex-post values.

(i) 'voluntary exchange': $c^p \leqslant c$, $i^p \leqslant i$, $g^p \leqslant g$, $y^p \leqslant y$, $m^p \leqslant m^s$, $m^p \leqslant m^d$.

(The superscript p refers to ex-post values)

(ii) 'perfect market': $m^p = \min\{m^s, m^d\}$
$y^p = \min\{y, c+i+g\}$
(iii) 'applicability of the theory': $c+i+g=y$
$m^s = m^d$

Taken together these three principles assert that ex-ante and ex-post values of the respective variables coincide: $c^p = c$, $i^p = i$, $g^p = g$, $y^p = y$, $m^p = m^s = m^d$.[5]

In our account of the labour market the principle of the perfect market is not needed in order to establish the equality of ex-ante and ex-post values.

(i) 'voluntary exchange': $w^d \leqslant w^p \leqslant w^s$
(iiia) 'applicability of the *classical* theory': $w^s = w^d$
(iiib) 'applicability of the *Keynesian* theory': $(w^d <)\ w^p = w^s$

If one wants to apply the classical theory in a particular application, then (i) and (iiia) imply that $w^p = w^s = w^d$. If one wants to apply the Keynesian theory, then (i) and (iiib) imply that $w^p = w^s\ (> w^d)$.

The above analysis implies that the respective equilibrium conditions are needed in order to justify the identification of ex-ante and ex-post values. In other words, the equilibrium theories are actually used in order to measure the demand and supply variables. In order to capture this idea I introduce the following definition: a term is *indirectly observed* with respect to theory T if it is *actually* measured by using (the validity of) T.[6] I conclude that the demand and supply variables are indirectly observed with respect to the macroeconomic equilibrium theories.[7]

Related to these points is the distinction between voluntary and involuntary unemployment. Unemployment is said to be voluntary if at the prevailing level of employment the wage rate firms are willing to offer is equal to the wage rate (employed) households demand. In this case *unemployed* agents demand a too high wage to get a job. Unemployment is said to be involuntary if wages are equal to the wage rate firms are willing to offer and higher than the wage rate households demand. In this case unemployed persons do not demand a too high wage rate, but they simply can not get a job.

In the previous sections I have shown that the structures of classical and Keynesian theories are almost the same. The only important difference reveals itself in the equilibrium condition of the labour market. The analysis of this section shows that the equilibrium conditions are used to 'measure' supply and demand. It thus turned out that classical and Keynesian economists interpret statistics of the product and money market in the same way, while they differ with respect to the interpretation of labour market statistics, i.e., they differ with respect to the question whether unemployment is voluntary or not. As long as questionnaires are not regarded as proper measuring methods of demand and supply there is no way to determine which interpretation is the most proper one: classical and Keynesian interpretation are valid as long as they remain within the closure of their own theories.[8] Thus, at the present stage of applied economics the distinction between voluntary and involuntary unemployment is not made on empirical grounds.[9]

Finally, the status of the state values has to be discussed. Are the state values directly or indirectly observed? All demand and supply functions have to be understood as desired actions at the prevailing value of the state values. The state values are taken as given by the agents. Hence, it is natural to view the state values as directly observed. If the demand and supply variables are identified according to the classical theory, then the state values fulfil automatically the equilibrium conditions of the classical theory. On the other hand if the demand and supply variables are identified in the Keynesian way, then the state values fulfil automatically the equilibrium conditions of the Keynesian theory.

NOTES

* The author thanks W. Balzer, B. Hamminga, T. Kuipers and G. Pikkemaat for critical remarks on an earlier version of this paper. Responsibility for any remaining errors is his own.

1 I do not claim that these definitions exhaust the category of definitions of the notion of an equilibrium. The only claim I want to make is that one can encounter these definitions in economics and that the definitions play a role in understanding the macro-theories presented in this paper.

2 An example of a 'situation on the two markets' is the following: excess demand on the product market and excess supply on the money market.

3 Remark that the 'inverse' notation is a nice way of presenting the distinction between voluntary and involuntary unemployment. See also Section 6.

4 Hands introduces the equilibrium condition as a specialization. However, if the axioms of the model are fulfilled one can simply prove that an equilibrium price vector exists.

Thus, also in Hands' reconstruction equilibrium has to be seen as a part of the model and not as a specialization.

[5] The reader is reminded of the fact that $c^p + i^p + g^p \equiv y^p$ holds by definition.

[6] Note that this definition resembles the structuralistic definition of theoretical terms in which a term is said to be T-theoretical if there is no way to measure it independently of theory T. So, a term is theoretical if it *can* not be measured independently of theory T, while a term is indirectly observed if it *is* not measured independently of theory T. An indirectly observed term can be observable (i.e., non-theoretical), while a theoretical term can not be directly observed.

[7] Note that the equilibrium conditions are used *only* in a conceptual way. The equilibrium axioms determine the setting, the interpretative framework in which the ex-post observations have to be identified with their ex-ante counterparts in order to apply the theory.

[8] Haslinger (1982) regards involuntary unemployment as theoretical in the structuralistic sense. I have posed that both voluntary *and* involuntary unemployment are indirectly observed; see also Note 5.

[9] I do not discuss the question whether the macro-theories have empirical content or not. The reader is referred to Janssen and Kuipers (1988) who discuss this question with respect to general equilibrium theory. Their analysis covers to a large extent the issues involved in macroeconomics.

REFERENCES

Arrow, K. J.: 1968, 'Economic Equilibrium', in the *International Encyclopedia of the Social Sciences*, MacMillan and Free Press, London, 376–89.

Balzer, W.: 1982, 'A Logical Reconstruction of Pure Exchange Economics', *Erkenntnis* **17**, 23–46.

Bénassy, J.-P.: 1982, *The Economics of Market Disequilibrium*, Academic Press, New York.

Branson, W. H.: 1979, *Macroeconomic Theory and Policy*, Harper and Row, New York.

Fomby, T., R. Hill and S. Johnson: 1984, *Advanced Econometric Methods*, Springer, Berlin.

Frisch, R.: 1936, 'On the Notion of Equilibrium and Disequilibrium', *Review of Economic Studies* **3**, 100–5.

Hands, D.: 1985a, 'The Logical Reconstruction of Pure Exchange Economics: Another Alternative', *Theory and Decision* **19**, 259–78.

Haslinger, F.: 1982, 'Structure and Problems of Equilibrium and Disequilibrium Theory', in W. Stegmüller, W. Balzer and W. Spohn (eds.), *Philosophy of Economics*, Springer-Verlag, Berlin, 63–84.

Haslinger, F.: 1983, 'A Logical Reconstruction of Pure Exchange Economics: An Alternative View', *Erkenntnis* **18**, 115–29.

Hicks, J.: 1937, 'Mr. Keynes and the "Classics"; A Suggested Interpretation', *Econometrica* **5**, 147–59.

Janssen, M. and Kuipers, T.: 1988, 'Stratification of General Equilibrium Theory: a Synthesis of Reconstructions', this volume.

Machlup, F.: 1963, *Economics Semantics*, Prentice Hall, Englewood Cliffs.

Sargent, T.: 1979, *Macroeconomic Theory*, Academic Press, New York.
Solow, R. M.: 1980, 'On Theories of Unemployment', *Am. Ec. Review* **70**, 1–11.
Standing, G.: 1981, 'The Notion of Voluntary Unemployment', *International Labour Review* **120**, 563–79.
Weintraub, E. R.: 1985, *General Equilibrium Analysis: Studies in Appraisal*, Cambridge University Press, Cambridge.

Manuscript received 25 January 1988

Department of Econometrics
University of Groningen
P.O. Box 800
9700 AV Groningen
The Netherlands

MAARTEN C. W. JANSSEN AND THEO A. F. KUIPERS*

STRATIFICATION OF GENERAL EQUILIBRIUM THEORY: A SYNTHESIS OF RECONSTRUCTIONS

1. INTRODUCTION

A number of authors have already tried to reconstruct general equilibrium theory (GET) in a structuralist way, in particular, following the order of appearance, Händler (1980), Balzer (1982), Haslinger (1983) and Hands (1985a). However, in view of their mutual criticism, see the papers referred to and in addition Balzer (1985) and Hands (1985b), this has thus far not led to a generally accepted reconstruction.

In this paper we will present a new attempt. However, we will not introduce a new point of view which is incompatible with those already published. Instead we will argue that it is possible to reconciliate main points of the existing reconstructions, provided one is prepared to introduce a stratification between the individual and the collective level.

Leaving out production, as is usual in the earlier papers, this leads to reconstructions of what we call individual demand theory (IDT) and collective demand theory (CDT) and their mutual relation: Sections 2, 3 and 4 respectively. In Section 5 we compare our synthetic reconstruction with the earlier ones, with special emphasis on features we did not want to take over.

Beside overlooking a fruitful distinction of levels the controversy is certainly also caused by the fact that most authors tried to do justice, in dogmatic structuralist lines, to theoretical and applied economics at the same time. In this respect we have followed economists, who used to make a sharp distinction. Then it becomes plausible to start with theoretical economics. More precisely, Sections 2, 3 and 4 claim to give reconstructions of GET (without production), as far as it is a theory developed by theoretical economists. In Section 6 we elaborate on this point. We will argue that 'theoretical-GET' is another illustration of Hamminga's view on structure and development of economic theories. In line with this, we will point out an interesting partial analogy between the goals and methods of theoretical economists and natural scientists.

Erkenntnis **30** (1989) 183–205.

In Section 7 we try to reconstruct the intended applications and the empirical claim related to exchange prices and exchanged amounts as they occur in applied economics. It turns out that there is not a strong relation with collective demand theory (CDT). In Section 8 we formulate the theoretically possible intended applications and the empirical claim as they are suggested by CDT seen from a structuralist point of view. In Section 9 we do the same for individual demand theory (IDT), where we distinguish between exchange- and interview-applications. In Section 10 we sketch the main things to be done to prove the theorem that the individual (exchange-) claim of IDT implies the collective claim of CDT. Section 11 makes some final remarks.

We conclude this introduction with a general remark about the value of (undogmatic) structuralist reconstructions of theories in general and economic theories in particular. Reconstructions may occasionally suggest new lines of research, but this is not the first goal of reconstructing. An adequate reconstruction will, by definition, not produce new insights in the object of the discipline in question, i.e., it will not produce new theories, but as a rule it will produce more insight in the existing theories. The latter is the primary objective and its results may first of all be assumed to be useful in the didactics of the discipline, in particular in advanced textbooks. Of course, beside this, more insight in existing theories may, but need not, be helpful in solving conceptual and empirical problems as well as in the further development of these theories.

Like most other scientific theories, economic theories are seldom presented in a maximally transparent way. Concerning GET, we hope to increase its transparancy.

2. INDIVIDUAL DEMAND THEORY

We start with presenting the class of proper models of individual demand theory (IDT). Later we will define classes of *in*complete structures. Classes of definitions are always essentially mathematical conditions, followed by their intended meaning in the present context and possibly by some notation convention.

DEFINITION 1.1. $\langle D, G, e, u, d\rangle$ is a *model of individual demand theory* ($\in$ IDT) iff

(I.1)	D is a non-empty, finite set i: index over D	(set of individuals)
(I.2)	G is a non-empty, finite set j: index over G, n: size of G	(set of types of goods)
(I.3)	e: $D \times G \rightarrow \mathbf{R}_0^+$ e_{ij}: i's initial endowment of good j	(individual initial endowment)
(I.4)	u: $D \times (\mathbf{R}_0^+)^n \rightarrow \mathbf{R}$ $u_i[\underline{y}]$: i's utility of quantity vector $\underline{y} = \langle y_1, \ldots, y_j, \ldots, y_n \rangle$	(individual utility function) (comsumption bundle)
(I.5)	d: $D \times G \times (\mathbf{R}_0^+)^n \rightarrow \mathbf{R}$	(individual excess demand function)

$d_{ij}(\underline{q})$: i's excess demand (supply, if negative) of good j at price vector $\underline{q} = \langle q_1, \ldots, q_j, \ldots, q_n \rangle$

(I.6) *commodity constraint*:[1] for all i, j and $\underline{q}$: $e_{ij} + d_{ij}(\underline{q}) \geqq 0$
(one cannot supply more than one has)

(I.7) *budget constraint*: for all i and $\underline{q}$: $\Sigma_j q_j d_{ij}(\underline{q}) \leqq 0$
(nobody's total endowment value can increase (hence, nor decrease))

(I.8) *utility constraint*: u is continuous and strictly quasi-concave with respect to each type of good

(I.9) *utility maximization*: for all δ satisfying 5, 6 and 7 and for all i and $\underline{q}$:

$$u_i[e_{i1} + \delta_{i1}(\underline{q}), \ldots, e_{ij} + \delta_{ij}(\underline{q}), \ldots, e_{in} + \delta_{in}(\underline{q})]$$
$$\leqq u_i[e_{i1} + d_{i1}(\underline{q}), \ldots, e_{ij} + d_{ij}(\underline{q}), \ldots, e_{in} + d_{in}(\underline{q})]$$

Assuming that the reader is familiar with all ingredients of the definition, we restrict ourselves to some structural remarks. The first five axioms define the mathematical status of the five components. Axioms (I.6) and (I.7) impose further constraints on the demand function in relation to the initial endowment. Axiom (I.8) makes the utility function manageable. Finally, Axiom (I.9) relates demand, in combination with initial endowment, to utility in the standard way.

The different roles of these axioms suggest many possibilities of defining classes of incomplete structures. The following three classes are partly inspired by structuralist practice and partly by theorem directed considerations. According to standard structuralistic practice we define potential models as follows

DEFINITION 1.2. $\langle D, G, e, u, d\rangle$ is a *potential* model of IDT ($\in \mathrm{IDT}_p$) iff (I.1), (I.2), (I.3), (I.4) and (I.5) hold.

From the standard structuralist point of view it would now be plausible to define *partial* potential models by throwing the (relatively) theoretical components out of the potential models. However, at this point we do not want to deal with the question whether or not demand and/or utility are theoretical terms. In contrast, we want to focus on distinctions which give a clear view on the kind of theorems that are proved by theoretical economists. In the present context the following

DEFINITION 1.3. $\langle D, G, e, u\rangle$ is a *restricted* model of IDT ($\in \mathrm{IDT}'$) iff (I.1), (I.2), (I.3), (I.4) and (I.8) hold.

is pressing itself because of a well-known theorem which reads in our terms:

THEOREM 1. If $\langle D, G, e, u\rangle \in \mathrm{IDT}'$ then there is precisely one d such that $\langle D, G, e, u, d\rangle \in \mathrm{IDT}$.

In view of Definition 1.3 it is plausible to define, now again in structuralistic lines:

DEFINITION 1.4. $\langle D, G, e, u\rangle$ is a *potential restricted* model if IDT ($\in \mathrm{IDT}'_p$) iff (I.1), (I.2), (I.3) and (I.4) hold.

3. COLLECTIVE DEMAND THEORY

It is frequently said that GET is essentially a theory of markets, not of individuals. This section will illustrate this claim. Again we start with proper models.

DEFINITION 2.1. $\langle G, z, \underset{\sim}{p}\rangle$ is a *model of collective demand theory* (CDT) iff

(C.1)	G is a non-empty, finite set of size n	(set of types of goods)
(C.2)	$z\colon G \times (\mathbf{R}_0^+)^n \to \mathbf{R}$	(collective excess demand functions)
	z_j: excess demand of good j at price vector $\underset{\sim}{q}$.	
(C.3)	$\underset{\sim}{p} \in (R_0^+)^n$, $\underset{\sim}{p} \neq 0$	(equilibrium price vector).

(C.4) *continuity axiom*: z is continuous with respect to each price component.

(C.5) *proportionality axiom*: for all $\underset{\sim}{q} \in (\mathbf{R}^+)^n$ and $\alpha \in \mathbf{R}^+$ $z(q) = z(\alpha q)$. (Demand does not change if all prices multiply by a constant).

(C.6) *collective budget constraint*: for all $\underset{\sim}{q} \in (\mathbf{R}^+)^n$ $\Sigma_j q_j \times z_j(\underset{\sim}{q}) = 0$ (the value of the total endowment does not change, Walras' axiom).

(C.7) *equilibrium axiom*: for all j $z_j(\underset{\sim}{p}) = 0$ or $(z_j(\underset{\sim}{p}) < 0$ and $p_j = 0)^2$. (Demand equals supply, with room for excess supply of 'free goods').

The following definitions are motivated in the same way as Definition 1.2, 1.3 and 1.4.

DEFINITION 2.2. $\langle G, z, p\rangle$ is a *potential* model of CDT ($\in \text{CDT}_p$) iff (C.1), (C.2) and (C.3) hold.

DEFINITION 2.3. $\langle G, z\rangle$ is a *restricted* model of CDT ($\in \text{CDT}'$) iff (C.1), (C.2), (C.4), (C.5) and (C.6) hold.

DEFINITION 2.4. $\langle G, z\rangle$ is a *potential* restricted model of CDT ($\in \text{CDT}_p'$) iff (C.1) and (C.2) hold.

Again, 'restriction' has nothing to do with a distinction between theoretical and non-theoretical terms, but is directed to a theorem, in this case:

THEOREM 2. If $\langle G, z\rangle \in \text{CDT}'$ then there is at least one $\underset{\sim}{p} \in (\mathbf{R}_0^+)^n$ such that $\langle G, z, \underset{\sim}{p}\rangle \in \text{CDT}$.

This theorem states that restricted models of CDT are specific enough to guarantee the existence of a model of CDT. But a restricted model does not yet guarantee that a corresponding model is unique. To obtain (relative) uniqueness we have to specialize as follows

DEFINITION 3.1. $\langle G, z, \underset{\sim}{p}\rangle$ is a *special model of* CDT ($\in$ SCDT) iff

$\langle G, z, \underset{\sim}{p}\rangle \in$ CDT and

(C.8) *gross substitutability*: for all $\underset{\sim}{q} \in (\mathbb{R}_0^+)^n$ $\delta z_j(\underset{\sim}{q})/\delta q_k > 0$ for $j \neq k$. (Increase of price of some good favours demand of all others goods).

DEFINITION 3.2. $\langle G, z\rangle$ is a *restricted* special model of CDT ($\in$ SCDT′) iff $\langle G, z\rangle \in$ CDT′ and (C.8).

THEOREM 3. If $\langle G, z\rangle \in$ SCDT′ then there is, (in view of (C.5)) apart from a constant factor, precisely one $\underset{\sim}{p} \in (\mathbb{R}_0^+)^n$ such that $\langle G, z, \underset{\sim}{p}\rangle \in$ SCDT.

4. RELATION BETWEEN IDT AND CDT

The relation between IDT and CDT is easy to describe in terms of the following plausible bridge definition of individual and collective excess demand.

DEFINITION 4 (*aggregation bridge*). If $\langle D, G, d\rangle$ satisfies (I.1), (I.2) and (I.5) then the collective excess demand $z_j(\underset{\sim}{q})$ of good j at price vector $\underset{\sim}{q}$ is $\Sigma_i d_{ij}(\underset{\sim}{q})$, which gives rise to the *collective excess demand function* $z^{\langle D,G,d\rangle}$: $G \times (\mathbb{R}_0^+)^n \rightarrow \mathbb{R}$.

The following 'upward theorems' are both plausible and easy to prove.

THEOREM 4 (4^p). If $\langle D, G, e, u, d\rangle \in$ IDT (IDT$_p$) then $\langle G, z^{\langle D,G,d\rangle}\rangle$ $\in$ CDT′ (CDT$'_p$).

At this point it is interesting to recall that Theorem 2 stated that every restricted model of CDT can be extended with a price vector to a model of CDT. Theorem 4 states on its turn that all the assumptions concerning collective excess demand functions, which are mentioned in CDT′, can be derived from 'more fundamental' assumptions concerning individual behaviour. Thus, Theorems 2 and 4 together provide explicit illustrations of the 'tendency implicit in price theory, particularly in its mathematical versions, to deduce all properties of aggregate behaviour from assumptions about individual economic agents' (Arrow (1968, p. 382)).

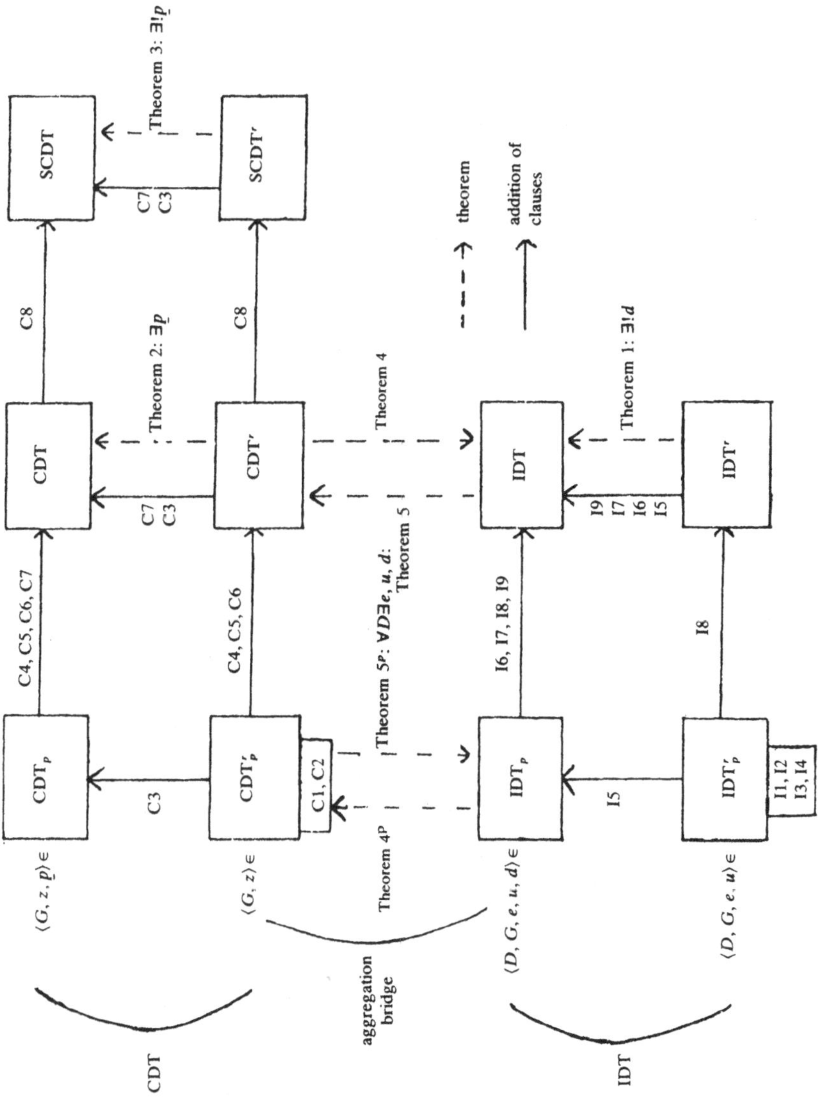

Fig. 1. A survey of individual and collective demand theory and their mutual relation.

Less trivial than the upward theorems are the 'downward theorems', at least the (restricted) model version, which was proved, among others, by Sonnenschein (1972).

THEOREM 5 (5^p). If $\langle G, Z, \rangle \in \mathrm{CDT}^r$ (CDT_p^r) then for all non-empty, finite d there are e, u, and d such that $z = z^{\langle D,G,d \rangle}$ and $\langle D, G, e, u, d \rangle \in \mathrm{IDT}$ (IDT_p).

In words, given a restricted model of *collective* demand theory it is always possible to think behind this model a model of *individual* demand theory for a given set of individuals.

Now we are in the position to give a complete survey of individual and collective demand theory, and their mutual relation: see Figure 1.

5. COMPARISONS TO EARLIER RECONSTRUCTIONS

In recent years different authors have proposed different ways to reconstruct general equilibrium theory without production; see e.g. Händler (1980), Balzer (1982, 1985), Haslinger (1983) and Hands (1985a). From the point of view of our two-level reconstruction we are able to put earlier controversies in a proper context. Despite some differences the reconstructions by Balzer and Haslinger share the emphasis on the individual part of GET. Hands, on the other hand, emphasized that GET is a theory about prices in competitive markets, although he mentions the underlying utility maximization in an indirect way by referring to the result of Debreu (1974), which is related to the result of Sonnenschein (1972), to which we referred in the context of Theorem 4. All three authors hold different opinions with respect to the question which concept of GET is the central and crucial one. Balzer (1982) argued that the utility function is the crucial concept of GET. According to Haslinger the individual demand function has to be considered as such, while Hands poses that the market excess demand function has to be viewed as the central and crucial concept. In our analysis everything is falling into its proper position. Utility is the central concept of IDT, the market excess demand function is the central concept of CDT, with the individual excess demand function mediating between them.

The second main topic in the earlier papers on reconstruction of

GET concerns the question whether market clearing (which is implied by our (C.7)) is a basic element of a model of GET or whether it has to be seen as a specialization. Balzer (1982) distinguishes between equilibrium distributions and market clearing. He sees market clearing as a specialization of GET, while maintaining that a set of equilibrium distributions, i.e., a set containing all distributions at which individuals maximize their utility, belongs to a model of GET. It is important to note that Balzer imposes two constraints on utility maximization: the usual budget constraint and a so-called consumption set, which asserts that collective demand can not be greater than the total amount existing in the economic system. Balzer's presentation with respect to this point is rather strange. (1) The second Theorem of Welfare Economics shows that every Pareto optimal state (roughly speaking a state at which all individuals are utility maximizers) can be supported by a market clearing situation. Market clearing is thus more or less implied by Balzer's equilibrium distributions, hence it cannot be treated as a specialization of GET. (2) Balzer's consumption set is quite similar to the requirement of market clearing (our Axiom (C.7)). He thus imposes a kind of 'market constraint' on utility maximizing behaviour of individuals so that it is (again) not clear why he treats market clearing also as a specialization. Balzer (1985)'s distinction between 'equilibrium$_{real}$' and 'equilibrium$_{get}$' will be discussed implicitly by treating the accounts of the market clearing condition by Haslinger and Hands.

Haslinger and Hands see the notion of an equilibrium in the context of GET as identical to the market clearing condition. Haslinger poses that market clearing is a basic element of GET, whereas Hands asserts that it is a specialization of GET. Haslinger (1983, p. 123) points out that 'equilibrium theorists . . . restrict their attention to Walrasian equilibrium states only' and that 'all laws of GET exclusively apply to this subset of states'. Hands (1985a, p. 213), on the other hand, stresses the importance of adjustment processes. In our point of view both positions can be reconciliated if a proper distinction is made between theoretical and more applied research with respect to GET. In the theoretical literature a good deal of discussion is concerned with adjustment processes and the characterization of the set of equilibrium states of the system. Results on adjustment processes are not very promising and we agree with Haslinger that economists have not much to say about such processes. In fact we have not treated specializations in which adjustment processes are analyzed. Existence proofs show that the set of equilibrium states of the system is not empty, i.e., under suitable assumptions one can show that the uncoordinated actions of self interested individuals need not (necessarily) lead to chaos. Thus, the non-emptiness of the set of equilibrium states of the system is central to GET and it is curious that Hands introduces the existence of an equilibrium set of prices as a specialization, because this existence is already implied by his axioms (D.1–5), (D.2–2) and (D.2–3) (in our reconstruction (C.4), (C.5) and (C.6)). However, the non-emptiness of the set of equilibrium states of the system does not guarantee at all that an actual state of an economy is an equilibrium state. Nevertheless, if GET is used in applied research actual (observed) states of the economy are usually identified as equilibrium states of the system. But this does not mean that the identification procedure is a central element of theoretical research with respect to GET. Thus, Haslinger is wrong in interpreting the equilibrium axioms of theoretical research as asserting that actual states of a concrete economy are equilibrium states. We elaborate further on the distinction between theoretical and applied research in the Sections 6 and 7.

6. THEORETICAL ECONOMICS: ITS GOALS AND METHODS

Hamminga (1982, 1983) claims that a main activity of theoretical economists is to prove so-called ‘interesting theorems’ on the basis of an ordered set of fixed and variable conditions. In particular, they aim to prove such a theorem for a growing number of conceptually possible situations by four strategies: field-extension, weakening of conditions, alternative conditions and conditions for conditions. He arrives at this tentative view on especially neo-classical economics by a case study of (the development of) the theory of international trade.

The presented partial analysis of GET highlights that GET is grist to the mill of Hamminga. In particular, Theorems 1, 2, 3, and 5 are examples of theorems which are certainly found to be interesting by economists. Moreover, there are many possibilities to illustrate the strategies. We will confine ourselves to one example.

The gross-substitutability specialization we have defined in Section 3 is the most well-known specialization under which uniqueness of the equilibrium price vector can be proved. Wald (1936) has established this result for the first time. However, Arrow and Hurwicz (1960), among others in the late fifties and in beginning of the sixties, have proved that uniqueness also holds under weak gross substitutability and connectedness, which is a clear case of 'weakening of conditions', because a gross substitutability economy is certainly connected.

Economists consider successful application of one of the strategies as (theoretical) progress. According to Hamminga they do so because the new result increases the plausibility that the interesting theorem is true in the actual world. This may be faithful to the actual intentions of economists, but we do not claim that these strategies are functional for these intentions.

Be this as it may, we like to draw attention to a formal analogy between progress in the natural sciences and the described type of theoretical progress in economics. The analogy is particularly pressing as far as the latter type of progress concerns existence and uniqueness theorems. Hence, we will restrict the description of the analogy to such theorems, leaving room for the plausible possibilty to extend the analogy to other types of interesting theorems.

The general situation may now be described as follows. Defined is a set of potential models M_p, a set of restricted potential model M_p^r, a restriction function ρ from M_p into M'_p and a set of models M, being a subset of M_p.

Existence and uniqueness theorems deal with a *scope* S, being a subset of M_p^r, about which they claim respectively:

(ET) (Existence Theorem): if $x \in s$ then $\exists y \in M\ \rho(y) = x$

(UT) (Uniqueness Theorem): if $x \in s$ and $y,\ y' \in M,\ \rho(y) = \rho(y') = x$ then $y = y'$

(EUT) (ET and UT); if $x \in s$ then $\exists^1 y \in M\ \rho(y) = x$.

Theorems 2 and 5 are clear examples of ET and Theorem 1 and 3 of EUT.

Of course, the ultimate targets of these theorems are the largest scopes for which the theorem can be proved. They are the crucial unknowns and research is directed to approach them. At this point the analogy becomes relevant. In Kuipers (1982) it is argued that the basic type of truth approximation in the natural sciences consists of attempts to characterize the unknown set X of empirical possibilities of the context. Now it is plausible to define that a theory, with class of models, S_2 is *progressive* with respect to a theory S_1 if and only if the symmetric difference between S_2 and X (defined as $(S_2 - X) \cup (X - S_2)$) is a proper subset of that between S_1 and X. For the (easy to check) special case that S_1 is a proper subset of S_2 ($S_1 \subset S_2 \subseteq X$), it is possible to speak of *cumulative progress*.

By analogy, economists claim (cumulative) progress under the same conditions, where X is now interpreted as a target set described above. However, economists are in one important respect in an advantageous position. In the natural sciences X is always primarily an empirically circumscribed set and, hence, it is never possible to check progress-claims in a straightforward way. In contrast, the target sets in theoretical economics are basically mathematically circumscribed sets. This has the indirect consequence that, as far as economists do not make mistakes in their proofs, they can be certain about claims of cumulative progress. For in that case it is only necessary to check two proofs, in addition to whether scope S_1 is indeed a proper subset of scope S_2.

7. APPLIED ECONOMICS

From the structuralist point of view a plausible question is what are the intended applications of IDT and CDT, and hence what are their empirical claims.

Because most applications are on the market level we start with CDT. Moreover, we will first try to answer the two questions in a pragmatic way, i.e., without bothering about the particular structuralist meaning of intended applications and the empirical claim, but as faithful as possible to what applied economics are doing.

The pragmatic definition of intended applications has to include time and collective income, which has the following background. One of the main problems associated with applying general equilibrium theory to concrete events is that there is no time involved in the theory, while concrete economies are constantly changing in time. Applied economists deal with this problem by relating all variables at the collective level to time. However, if nothing changes at the individual level, collective excess demand functions do not change either, and equilibrium prices and exchanged amounts are fixed in all distinguished time periods. One way to avoid this rather implausible consequence is to pose that initial endowments change from period to period. The operational counterpart of initial endowments at the collective level is the collective income, i.e., the money value of total initial endowments. (Distribution effects, i.e., effects arising from different distributions of total initial endowments over individuals, are thus neglected.) Moreover, applied economists used to treat collective income as exogenous relative to exchange prices and exchanged amounts.

In the light of this we will define the intended applications and the empirical claim of what we will call applied collective demand theory (ACDT).

For convenience we will generally assume that G is a finite set of types of goods, T is a finite, ordered set of time periods.

DEFINITION 5: $\langle T, G, y, p, f \rangle$ is *a possible sequence of intended applications of ACDT* iff

(1)	$y: T \to \mathbb{R}_0^+$	(collective income)
(2)	$p: T \to (R_0^+)^n$	(exchange prices)
(3)	$f: T \times G \to \mathbb{R}_0^+$	(exchange amounts).

It has to be said that the set of intended applications has a pragmatic nature. It is not the case that economists regard *all* tuples $\langle T, G, y, p, f \rangle$ as intended applications, but instead all intended applications have the form as outlined in Definition 5. Economists do not fix a set of intended applications of a theory. Instead they adapt an (extreme) instrumentalistic point of view by applying their theories to all kinds of situations and look 'how good they will do'. If a theory does not give an accurate picture of a situation, then economists do *not* conclude that the theory is falsified, but they simply pose that the theory is apparently not appropriate to deal with the situation at hand.

Now, applied economists do not apply CDT to the defined sequence of intended applications, but another theory, viz. a theory of which the models specify distinct demand and supply functions that are linear functions of normalized prices and collective income and that are moreover related in time.

DEFINITION 6. $\langle T, G, y, p, d, s \rangle$ is *a sequence of related models of* ACDT iff

(1)	$y: T \to \mathbb{R}_0^+$	(collective income)
(2)	$p: T \to (\mathbb{R}_0^+)^n$	(equilibrium prices)

(3) $d^3: T \times G \times (\mathbb{R}_0^+)^n \to \mathbb{R}_0^+$ (collective demand)
$s: T \times G \times (\mathbb{R}_0^+)^n \to \mathbb{R}_0^+$ (collective supply)

(4) there are coefficient vectors $\underline{\alpha}^d$, $\underline{\alpha}^s$, $\underline{\beta}^d$, $\underline{\beta}^s$, $\underline{\gamma}^d$ $\underline{\gamma}^s$, with $\beta_j^d - \beta_j^s > 0$ for all j, such that for all t, j and $\underline{q}$

$d_j^t(\underline{q}) = \alpha_j^d + \beta_j^d \cdot q_j / \Sigma_k q_k + \gamma_j^d \cdot y^t$

$s_j^t(\underline{q}) = \alpha_j^s + \beta_j^s \cdot q_j / \Sigma_k q_k + \gamma_j^s \cdot y^t$

(5) $d_j^t(p^t) = s_j^t(p^t)$ for all t and j

Applied economists do not claim that, in our terms, the intended applications fit perfectly into models with the related linear form specified in Definition 6.4, but they leave always room for disturbance terms. We will not write this out, but we will just hint upon this by talking about "approximately" in the following definition.

DEFINITION 7. *Empirical claim of* ACDT with respect to an actual sequence of intended applications of ACDT $\langle T, G, y, p, f\rangle$:

> there are demand and supply functions $\underset{\sim}{d}$ and s such that $\langle T, G, y, p, d, s\rangle$ is a sequence of approximately related models of ACDT such that for all t:
>
> $$f^t = d^t(p^t) \; (= s^t(p^t)).$$

It is clear that d and s are determined by using the claim: hence, in structuralist terms, d and s are treated as theoretical functions of ACDT, whereas y, p and f are non-theoretical with respect to ACDT. (See Janssen (1987) for some nuances with respect to this point. His remarks in the context of macroeconomics are also valid for GET.)

The intriguing question now is of course what this has to do with CDT. It is our impression that neither applied nor theoretical economists are very much interested in this question, whereas both sides could profit from an answer.

It is easy to check that clause 4 of Definition 6 is such that the excess demand function z^t (of course defined as $d^t - s^t$) satisfies (C.4), (C.5) and (C.8) of (S)CDT. Moreover, clause 5 guarantees that z^t satisfies (C.7) for p^t. However, although z^t satisfies (C.6) trivially for p^t, due to clause 5, z^t will not automatically satisfy (C.6) for all price-vectors, to say the least. By consequence, ACDT is certainly not just a specialization of CDT, and hence Theorem 2 and Theorem 3 cannot yet be used.

We do not know whether (C.6) together with the other CDT-conditions can be satisfied at all by models of ACDT in the sense of Definition 6. We only know of functions satisfying all CDT-conditions, including (C.6), which do clearly not satisfy the specific linearity condition 4 of Definition 6. The question of the joint satisfiability does not seem to interest theoretical economists very much, and applied economists do not seem to bother about it. In our opinion, however, it is an intriguing question for further research.

8. A PLAUSIBLE CLAIM OF CDT

Let us now forget about standard applied economics and concentrate on the intended applications and empirical claims as they are suggested by CDT and IDT if one looks from a structuralist point of view.

We start with CDT and we will immediately focus on special models: SCDT.

DEFINITION 8. $\langle T, G, p, f\rangle$ is *a possible sequence of intended applications of SCDT* iff

(1) p: $T \to (\mathbb{R}_0^+)^n$ (exchange prices)
(2) f: $T \times G \to \mathbb{R}_0^+$ (exchanged amounts)

DEFINITION 9. $\langle T, G, d, s\rangle$ is *a sequence of d/s-models of* SCDT′ iff

(1) d/s: $T \times G \times (\mathbb{R}_0^+)^n \to \mathbb{R}$ (collective demand/supply)
(2) $\langle G, d^t - s^t\rangle \in$ SCDT′ (i.e., (C.1, 2, 4, 5, 6, 8) (restricted special models of CDT)

DEFINITION 10. $\langle T, G, d, s, p\rangle$ is *a sequence of d/s-models of* SCDT iff

(1) d/s: $T \times G \times (\mathbb{R}_0^+)^n \to \mathbb{R}$ (collective demand/supply)
(2) p: $T \to (\mathbb{R}_0^+)^n$ (equilibrium prices)
(3) $\langle G, d^t - s^t, p^t\rangle \in$ SCDT (i.e., + (C.3, 7) (special models of CDT)

It is plausible to call $d^t(p^t)$ of Definition 10 (collective) equilibrium demand and $s^t(p^t)$ (collective) equilibrium supply. Theorem 3 now implies that a sequence $\langle T, G, d, s\rangle$ satisfying Definition 9 generates a unique sequence $\langle T, G, d, s, p_*\rangle$ satisfying Definition 10. This will be used in the claim of SCDT with respect to an actual sequence of intended applications: the goods are exchanged at equilibrium prices in the corresponding equilibrium amounts generated by some sequence of restricted d/s-models. Formally:

DEFINITION 11. *Empirical claim of* SCDT with respect to an actual sequence of intended applications of SCDT $\langle T, G, p, f\rangle$:

> there are demand and supply functions d and s such that $\langle T, G, d, s\rangle$ is a sequence of d/s-models of SCDT′ such that the (according to Theorem 3) generated unique sequence of d/s-models of SCDT $\langle T, G, d, s, p_*\rangle$ is such that for all t
>
> $$p^t = p_*^t$$
> $$f^t = d^t(p_*^t)\ (= s^t(p_*^t))$$

Like in Section 7, it is clear that d and s are determined by using the claim: hence, d and s are treated as theoretical functions of SCDT, whereas p and f are clearly non-theoretical with respect to SCDT. Finally, p_* is equated with p by the claim of SCDT and hence it is theoretical with respect to SCDT. However, it may also be said to be "less theoretical than" d and s, for p_* is, in combination with f, certainly not enough to determine d and s, whereas d and s determine p_*, due to Theorem 3, uniquely.

We conclude this section by noting that a sequence $\langle T, G, y, p, d, s\rangle$ of related models of ACDT (Definition 6) will give rise to a sequence $\langle T, G, p, d, s\rangle$ of d/s-models of SCDT, as soon as $d^t - s^t$ satisfies the collective budget constraint (C.6) (Walras' axiom) for all price vectors. As already mentioned at the end of Section 7, we do not know whether all the relevant conditions can be satisfied together.

9. PLAUSIBLE CLAIMS OF IDT

In the same spirit as in Section 8 we will now formulate two possible empirical claims of IDT. D will generally refer to a finite set of individuals.

We start with a claim about (excess) demand as one might registrate by interviewing people.

DEFINITION 12. $\langle T, D, G, e, d\rangle$ is *a possible sequence of intended interview-applications of* IDT iff

(1)	$e\colon T \times D \times G \to \mathbb{R}_0^+$	("initial" endowment)
(2)	$d\colon T \times D \times G \times (\mathbb{R}_0^+)^n \to \mathbb{R}$	(stated excess demand)
(3)	for all t, i, j and $\underline{q}$: $e_{ij}^t + d_{ij}^t(\underline{q}) \geqq 0$	(commodity constraint)
(4)	for all t, i and $\underline{q}$: $\Sigma_j q_j d_{ij}^t(\underline{q}) = 0$	(budget constraint)
(5)	for all t, j and $\underline{q}$: $\Sigma_i d_{ij}^t(\underline{q}) = 0$	(cleared markets)

DEFINITION 13. $\langle T, D, G, e, u\rangle$ is *a sequence of related models of* IDT′ iff

(1)	$e\colon T \times D \times G \to \mathbb{R}_0^+$	(initial endowment)
(2)	$u\colon D \times (\mathbb{R}_0^+)^n \to \mathbb{R}$	(time-invariant(!) utility)
(3)	$\langle D, G, e^t, u\rangle \in$ IDT′ (i.e., (I.1, 2, 3, 4, 8))	(restricted models of IDT)

DEFINITION 14. $\langle T, D, G, e, u, d\rangle$ is *a sequence of related model of* IDT iff

(1)	e as in Definition 13.1	
(2)	u as in Definition 13.2	
(3)	$d\colon T \times D \times G \times (\mathbb{R}_0^+)^n \to \mathbb{R}$	(utility maximizing individual excess demand)
(4)	$\langle D, G, e^t, u, d^t\rangle \in$ IDT (i.e., + (I.5, 6, 7, 9))	(models of IDT)

Theorem 1 implies that a sequence $\langle T, D, G, e, u\rangle$ satisfying Definition 13 generates a unique sequence $\langle T, D, G, e, u, d_*\rangle$ satisfying Definition 14. This will be used in the claim of IDT with respect to an actual sequence of interview-applications: the stated excess demand is the utility maximizing individual excess demand generated by some sequence of restricted models based on a time-invariant utility function. Formally:

DEFINITION 15. *Empirical claim of IDT* with respect to an actual sequence of intended *interview*-applications of IDT $\langle T, D, G, e, d\rangle$:

there is a utility function u such that $\langle T, D, G, e, u\rangle$ is a sequence of related models of IDT′ such that the (according to Theorem 1) generated unique sequence of related models of IDT $\langle T, D, G, e, u, d_*\rangle$ such that for all t

$$d^t = d_*^t$$

For analogous considerations about determination as in Section 8, assuming Definition 15, e and d are treated as non-theoretical with respect to IDT, and u and d_* are theoretical, and u is even more theoretical than d_*. Note that the situation is more complicated if we realize the fact that d can only be obtained by extra- and interpolation of finite 'interview-data'. However, the epistemic hierarchy of functions does not seem to be affected by this. Note too that there is no inconsistency between asserting that d and s are *treated* as theoretical functions with respect to ACDT and SCDT and, at the same time, asserting that d is *treated* as non-theoretical with respect to interview applications of IDT. We do not discuss whether d *is* (an 'ontological is') a theoretical function or not.

A second theoretically possible empirical claim of IDT constitutes the individual counterpart of the "exchange-claim" of CDT formulated in Section 8.

DEFINITION 16. $\langle T, D, G, e, p, g\rangle$ is *a possible sequence of intended exchange-applications of* IDT iff

(1)	e: $T \times D \times G \to \mathbb{R}_0^+$	(initial endowment)
(2)	p: $T \times G \to \mathbb{R}_0^+$	(exchange prices)
(3)	g: $T \times D \times G \to \mathbb{R}$	(exchanged amounts)
(4)	for all t, i, j: $e_{ij}^t + g_{ij}^t \geq 0$	(commodity constraint)
(5)	for all t, i: $\Sigma_i p_j^t g_{ij}^t = 0$	(budget constraint)
(6)	for all t, j: $\Sigma_i g_{ij}^t = 0$	(cleared markets)

Now the claim of IDT with respect to an actual sequence of exchange applications reads of course: the exchanged amounts correspond to the utility maximizing individual excess demand at the exchange prices generated by some sequence of restricted models based on a time-invariant utility function. Formally:

DEFINITION 17. *Empirical claim of* IDT with respect to an actual sequence of intended *exchange*-applications of IDT $\langle T, D, G, e, p, g\rangle$:

> there is a utility function u such that $\langle T, D, G, e, u\rangle$ is a sequence of related models of IDT$'$ such that the (according to Theorem 1) generated unique sequence of related models of IDT $\langle T, D, G, e, u, d_*\rangle$ is such that
>
> $$g^t = d_*^t(p^t)$$

Note that this is a rather strong claim: it asserts, among other things, that actual prices are equilibrium prices, in the sense that at the actual prices all individuals are able to fulfil their desired transactions (see also Section 10).

Plausible determination considerations lead now to the conclusion that, assuming Definition 17, e, p and g are non-theoretical with respect to IDT, that u and d_* are theoretical, and that d_*, though not fully determined by g^t, is nevertheless less theoretical than u, for the latter determines the former (according to Theorem 1) whereas the converse does not hold.

It is interesting to remark that Definitions 15 and 17 can be combined. The combined claim asserts that there is a utility function with corresponding demand function d_* such that $g^t = d_*^t = d^t$. However, this combined claim leads to inconsistencies when $g^t \neq d^t$, i.e., when agents do not respond honestly to questionnaires. The uncertainty with respect to the reliability of questionnaires is precisely the reason why economists hesitate to treat d at the collective level (see ACDT) as a non-theoretical term.

10. MICRO-REDUCTION OF THE COLLECTIVE CLAIM

A plausible last question is the relation between the individual and collective exchange claims. Roughly speaking, one, in particular a theoretical economist, may and will expect that the individual claim implies the collective claim. We will formulate the precise implication for SCDT.

On the individual level we can also formulate *special* models of IDT (SIDT), i.e., models of IDT of which the individual excess demand function satisfies:

(I.10) *substitutability*: for all i, j, $\underline{q}$ and $k \neq j$, $\partial d_{ij}(\underline{q})/\partial q_k > 0$.

By a trick we bypass the question of a sufficient condition on restricted models of IDT guaranteeing substitutability for the (according to Theorem 1) generated unique model. We just define *restricted* special models of IDT, SIDT′, implicitly as those restricted models which generate special models. Finally, the definition of the *empirical claim of SIDT* results of course from replacing in Definition 17 IDT by SIDT and IDT′ by SIDT′.

Later it will be used that Theorem 4 can be reformulated with an intermediate stage: a model of IDT generates a *restricted d/s-model* of CDT, by first aggregating the positive individual demands separately from the non-positive individual demands, i.e., the positive individual supplies. Here a restricted *d/s*-model is of course by definition such that $z =_{df} d - s$ generates a restricted model of CDT. It is now also easy to check that Theorem 4 can be strengthened to the statement that a *special* model of IDT generates (a restricted *d/s*-model of CDT that generates) a restricted *special* model of CDT, i.e., a restricted model of CDT, satisfying gross substitutability (C.8).

To formulate the collective claim implied by the empirical claim of SIDT it is finally necessary to note that a possible sequence $\langle T, D, G, e, p, g\rangle$ of intended exchange-applications of IDT (Definition 16), hence of SIDT, generates a unique possible sequence $\langle T, G, p, f\rangle$ of intended applications of SCDT (Definition 8) by equating f with the sum of the positive exchanged amounts (i.e., the amounts bought by individuals). Note that this sum is, due to the assumed cleared markets, equal to the sum of the sold amounts.

Now the following theorem is not difficult to prove on the basis of the informal definitions, the elaboration of Theorem 4 and Definitions 8 and 16.

THEOREM 6. The empirical claim of SIDT with respect to an actual sequence of intended exchange-applications of SIDT implies the empirical claim of SCDT with respect to the generated sequence of intended applications of SCDT.

It is tempting to make the following terminological remark. Theorem 6 states that a certain collective claim can be deductively derived from a claim made by an individualistic (or micro-) theory. According to a recent diagnosis of terms used in philosophy of science (Kuipers, 1987) this cannot only generally be called a deductive explanation of a collective claim, due to the apparent occurrence of an application step of a micro-theory and an aggregation step, and the absence of a transformation and an approximation step, it should more precisely be called: a deductive homogeneous micro-reduction of the collective claim.

Besides the remarks on the reduction of the collective claim to the individual claim we have to pay attention to the reversed relation, because – as is already mentioned – the individual exchange claim is theoretically possible, but not often encountered in economic analysis (see below). Economists treat the individual and the collective claim as equivalent: not only is the collective claim implied by the individual claim (as the foregoing remarks on micro-reduction make clear), but the individual claim is also seen to be implied by the collective claim. This last implication hinges on two plausible (and often made) assumptions: (i) the "principle of voluntary exchange", which says that individuals cannot be enforced to trade more than they like to and (ii) the "principle of perfect markets", sometimes referred to as the "short-rule", which says that there cannot be unsatisfied demand and supply on the same market. (See Janssen (1987) for the use macro-economic theories make of these principles in interpreting economic statistics.)

In line with these remarks economists actually adhere to a sophisticated version of the individual claim. Instead of claiming that all individuals are able to carry out their most preferred actions (see Definition 17), one should interpret the individual claim economists are adhering to as a conditional one: if there is a market excess demand (supply), then all potential sellers (buyers) are able to get what they supply (demand). (Remark that this is another way to state the "principle of perfect markets".) It is clear that the collective claim cannot be reduced to this sophisticated version of the individual claim. As this sophisticated individual claim is more in line with the way economists interpret their theories we doubt whether present economic theories can be seen as elements of a reductive research programme as is commonly alleged.

11. FINAL REMARKS

Although the "definition-density" in this paper, in particular in the last sections, is indeed very high, we believe to have provided more insight in the structure of GET and in the possibilities to make empirical claims with GET.

The present analysis raises many new questions, some within the present account of GET without production, others crossing this boundary. Of course, the main question of the latter type is how the present analysis can be extended to include production. We hope and we believe that the structuralist start made by Hamminga and Balzer (1986) can be adapted in a direction which is more in line with the present stratified approach of demand theory. But this can only be the subject of a new paper.

NOTES

* The authors would like to thank F. Haslinger for his constructive criticism.

[1] Although there may be peculiar interpretations of GET for which this axiom does not hold, it is a conceptual requisite in order to be able to talk about utility maximization, because the utility function is defined only for nonnegative amounts of goods. It is for this reason that it is included in the definition.

[2] If there is at least one individual who is non-satiated at every consumption bundle, then all goods which can be obtained at zero prices are in excess demand. In this case the equilibrium axiom amounts to saying that demand and supply balance on all markets. Thus, the case between brackets is exceptional.

[3] "d" should not be confused with the individual excess demand function in Definition 1.1.

REFERENCES

Arrow, K. J.: 1968, 'Economic Equilibrium', *International Encyclopedia of the Social Sciences*, MacMillan and Free Press, pp. 376–89.

Arrow, K. J. and L. Hurwicz: 1960, 'Competitive Stability and Weak Gross Substitutability', *International Economic Review* **1**, 38–49.

Balzer, W.: 1982, 'A Logical Reconstruction of Pure Exchange Economics', *Erkenntnis* **17**, 23–46.

Balzer, W.: 1985, 'The Proper Reconstruction of Exchange Economics', *Erkenntnis* **23**, 185–200.

Debreu, G.: 1974, 'Excess Demand Functions', *J. of Mathematical Economics* **1**, 15–21.

Hamminga, B.: 1982, 'Neoclassical Theory Structure and Theory Development: The Ohlin Samuelson Programme in the Theory of International Trade' in W. Stegmüller et al. (eds.), *Philosophy of Economics*, Springer, Berlin, pp. 1–16.

Hamminga, B.: 1983, *Neoclassical Theory Structure and Theory Development*, Springer, Berlin.

Hamminga, B. and W. Balzer: 1986, 'The Basic Structure of Neoclassical General Equilibrium Theory', *Erkenntnis* **25**, 31–46.

Händler, E.: 1980, 'The Logical Structure of Modern Neoclassical Static Microeconomic Equilibrium Theory', *Erkenntnis* **15**, 33–53.
Hands, D.: 1985a, 'The Logical Reconstruction of Pure Exchange Economics: Another Alternative', *Theory and Decision*, 259–78.
Hands, D.: 1985b, 'The Structuralistic View of Economic Theories: A Review Essay', *Economics and Philosophy* **1**, 303–35.
Haslinger, F.: 1983, 'A Logical Reconstruction of Pure Exchange Economics: An Alternative View', *Erkenntnis* **18**, 115–29.
Janssen, M.: 1987, 'Structuralist Reconstructions of Classical and Keynesian Macroeconomics', this volume.
Kuipers, T.: 1982, 'Approaching Descriptive and Theoretical Truth', *Erkenntnis* **18**, 343–378.
Kuipers, T.: 1987, 'A Decomposition Model for Explanation and Reduction', Abstract-LMPS-VIII, Section 6, Volume 4, Moscow, 1987, 328–37.
Sonnenschein, H.: 1972, 'Market Excess Demand Functions', *Econometrica* **40**, 549–63.
Wald, A.: 1936, 'Über einige Gleichungssysteme der Mathematischen Ökonomie', *Zeitschrift für Nationalökonomie*, 637–70.

Manuscript received 25 January 1988

University of Groningen
Department of Economics
P.O. Box 800
NL-9700 AV Groningen
The Netherlands

University of Groningen
Department of Philosophy
Westersingel 19
NL-9718 CL Groningen
The Netherlands

JOSEPH D. SNEED

MICRO-ECONOMIC MODELS OF PROBLEM CHOICE IN BASIC SCIENCE*

ABSTRACT. This paper describes the way in which a certain representation of basic scientific knowledge can be coupled with traditional microeconomic analysis to provide an analysis of "rational" research planning or "agenda setting" in basic science. Research planning is conceived as a resource allocation decision in which resources are being allocated to activities directed towards the solution of "basic scientific problems". A "structuralist" representation of scientific knowledge is employed to provide a relatively precise characterization of a basic scientific problem.

INTRODUCTION

In this paper I will describe the way in which a certain representation of basic scientific knowledge can be coupled with traditional micro-economic analysis to provide an analysis of "rational" research planning or "agenda setting" in basic science. Research planning will be conceived as a resource allocation decision in which resources are being allocated to activities directed toward the solution of "basic scientific problems". A "structuralist" representation of scientific knowledge will be employed to provide a relatively precise characterization of a basic scientific problem. The main thrust of the analysis will consist in describing the various ways in which values enter into these decisions. In particular, some effort will be devoted to distinguishing "internal" and "external" values operating in these decisions. The discussion will focus on exploiting this representation of scientific knowledge to say something about the "intrinsic scientific value" – "scientific value" for short – of solving certain basic scientific problems. This work is a part of a larger project whose aim is to describe the more general and typical situation of research planning in which technological and other external values play a role.

PROBLEMS IN BASIC SCIENCE

The fundamental concept I will use to analyse resource allocation in basic science is the concept of a "scientific problem". Roughly,

Erkenntnis **30** (1989) 207–224.

decision makers in basic science – individual scientists, research teams, research administrators – allocate resources to attempts to solve scientific problems. A micro-economic approach to this decision problem suggests that it may be fully understood in terms of values (or utilities) and probabilities attached to the solution of the problems by the decision makers.

My aim here is ultimately to clarify the nature of the values that might adhere to the solution of scientific problems. This aim is, I think, best approached by considering first how one might characterize a basic scientific problem. The account I have to offer of basic scientific problems is derived from a general picture (or representation scheme) for scientific knowledge. This scheme is described in detail in [1].

A REPRESENTATION OF SCIENTIFIC KNOWLEDGE

We may represent basic scientific knowledge generally as a net (directed graph) consisting of "theory elements" (nodes) linked together by intertheoretical links (arcs). Intuitively, theory elements are the smallest units of scientific theory that may be used alone to say something intelligible. They are the elementary building blocks out of which more complicated pieces of scientific theory are constructed. For example, Newton's theory of the gravitational force corresponds to a theory element which is one part of the full "theory" of Newtonian particle mechanics. The links between theory elements are intuitively paths permitting certain kinds of information to be passed between the elementary building blocks of scientific theory. The direction of the link indicates the direction of information flow.

Typically, each theory element **T** is linked to other theory elements by links going in two directions. That is, **T** receives information *from* other theory elements and, in turn, passes information *to* other theory elements. Links may even go both ways between theory elements. That is theory elements $\mathbf{T}_1$ and $\mathbf{T}_2$ may *exchange* information. For simplicity, I ignore this possibility here.

Theory elements are the intellectual focus of the social activity of scientific problem solving. Scientific communities may be organized around theory elements and/or collections of closely linked, "neighboring" theory elements – "theories" in one common sense of the word. These theories set the problems for the community as well as provide the means and criteria for the solution of these problems. "Theories" in this sense – e.g., Newtonian particle mechanics – play a role somewhat analogous to "paradigms" in Kuhn's account of "normal science" [5]. In the sociological literature on "theory choice" [6, 12], it appears that the "theories" that are being chosen are more often to be represented as our theory elements than as "theories" in the larger sense. Choice among theories in the larger sense appears, at least in "mature" areas of science, to be confined to those periods of scientific activity that may be viewed as truly "revolutionary" in Kuhn's original sense. For present purposes, it is necessary to see only a bit more precisely how this works. Those interested in a detailed, formal account may consult [1].

Each theory element **T** consists of a "conceptual core" **K** and a range of "intended applications" **I**. It is the range of intended applications that provides the most familiar kind of scientific problem and, as well, the key to characterizing other kinds. The range of intended applications consists of those problems that scientists committed to using **T** recognize as "soluble". The basis of this recognition has been focus of some attention in the sociology of science [2, 4, 6, 8, 11, 12]. For the moment, it suffices to say that our representation of scientific knowledge locates the considerations relevant to this choice in the configuration of links among theory elements. We will see below more precisely how this works. It is however clear already that our account is consistent with the "methodological dogma" that "... scientists define some problems as pertinent, and others as uninteresting or even illegitimate, primarily on the basis of theoretical commitments and other assumption structures" ([12], p. 74).

According to the knowledge representation scheme under consideration, the aim of those practicing science with **T** is to use **K** to make some claim about the whole of **I** which is generally not reducible to a conjunction of claims about individual members of **I**. This claim is "holistic" in a minimal sense just in that it is irreducibly about all of **I**. The typical "problem situation" associated with a theory element $\mathbf{T} = \langle \mathbf{K}, \mathbf{I} \rangle$ is that this claim will have been shown to be true for some finite sub-class of **I**, call it **A**. "Open problems" are then members of $\mathbf{I} \sim \mathbf{A}$. That is, an open problem is showing that a member of $\mathbf{I} \sim \mathbf{A}$ can be added to A preserving the truth of the theory element's claim for the augmented set. "Problem choice" is then – in the simplest case – just choosing which members of $\mathbf{I} \sim \mathbf{A}$ work on.

The intended applications $\mathbf{I}$ of a theory element $\mathbf{T} = \langle \mathbf{K}, \mathbf{I} \rangle$ are determined by the links running *to* $\mathbf{T}$ from other theory elements. Intuitively, these links provide the data about which $\mathbf{T}$ is a theory. In the language of traditional logical empiricism, they provide empirical interpretation for some parts of the vocabulary of $\mathbf{K}$. In many cases, other theories $\mathbf{T}'$ with links going *to* $\mathbf{T}$ may be viewed as providing means of measuring the values of empirical quantities in $\mathbf{K}$. In these cases, $\mathbf{T}'$ may be viewed as a theory about the measurement of these quantities. More generally, these links provide the context for the practice of science with theory element $\mathbf{T}$. They link this practice with other parts of science. Developments in other parts of science – e.g., discoveries that result in new instruments and/or new methods of observing certain things impinge on the practice of $\mathbf{T}$-science via these links. In this way, the links determine the intended applications $\mathbf{I}$ that are a part of practicing $\mathbf{T}$-science.

This representation of the intended applications of $\mathbf{T}$ will, in many specific cases, do some violence to our ordinary usage of the word 'theory'. The kinds of knowledge that legitimate the results of measurement and observation will frequently be more accurately viewed as "common sense" rather than "theoretical" knowledge. In some cases this common sense knowledge may be viewed as providing the intellectual focus of a technological or experimental tradition. This fact occasions only linguistic abuse, provided that "common sense" knowledge can be represented with the same formal tools as more explicit "theoretical" knowledge. At this time, whether formal representation of common sense knowledge is possible with the tools sketched here is an open question. We consider briefly below the possibility of formalizing the knowledge embodied in technological traditions. See, however [3].

Problem solving with $\mathbf{T}$ works in such a way that successfully solved problems in $\mathbf{T}$ provide part of the description of the intended applications $\mathbf{I}''$ of theories $\mathbf{T}'' = \langle \mathbf{K}'', \mathbf{I}'' \rangle$ with links going *from* $\mathbf{T}$ to $\mathbf{T}''$. Thus, problem solving with $\mathbf{T}$ may be seen as a kind of transformation or filter on information. Information coming into $\mathbf{T}$ from theory elements $\mathbf{T}'_1 \ldots, \mathbf{T}'_n$ is "filtered" into the results of successfully solved problems and then passed on *to* other theories $\mathbf{T}''_1 \ldots, \mathbf{T}''_m$ to partially determine their intended applications $\mathbf{I}''_1 \ldots, \mathbf{I}''_m$.

To understand the kind of information processing that counts as problem solving with **T** we must be more precise about the intended applications **I** and the claim that **T** makes about them. To do this, I must first say more about what is in the conceptual core **K** of a theory element. A core **K** consists of four parts:

$\mathbf{M_p}$ the vocabulary or **conceptual apparatus** characteristic of the theory element;

M an **empirical law** or systematization formulated in the vocabulary stipulated by $\mathbf{M_p}$;

It is convenient to think of $\mathbf{M_p}$ as the class of all systems that may be described using the vocabulary characteristic of the theory element and **M** as a sub-class of these structures satisfying some specific law. Both $\mathbf{M_p}$ and **M** are represented formally as classes of "models" in the mathematical logician's sense.

C some conditions or **constraints,** analogous to the laws **M**, that are imposed on collections of systems from $\mathbf{M_p}$, rather than on individual members of $\mathbf{M_p}$.

Intuitively, **K** contains means of imposing restrictions both on individual members of $\mathbf{M_p}$ and on collections of members of $\mathbf{M_p}$. These latter conditions entail that problem solving with **K** will generally have certain "holistic" features which are important for understanding problem choice. Formally, **C** may be conceived as a class of sub-classes of $\mathbf{M_p}$ satisfying some specific conditions on such sub-classes.

$\mathbf{M_{pp}}$ the vocabulary of the theory element that is linked to other theory elements via "interpreting links" going *to T* – the **non-theoretical conceptual apparatus.**

Formally, $\mathbf{M_{pp}}$ is also conceived as a class of systems that are all "fragments" (or sub-systems) of the systems in $\mathbf{M_p}$. Intuitively, these are the systems about which other theory elements may "possibly" provide information via links that "interpret" (or assign meaning to) the concepts that comprise them. 'Possibly' means here that there is a purely formal (conceptual or definitional) link between the concepts in $\mathbf{M_{pp}}$ and those in $\mathbf{M_p}$ in some other theory. The concepts that appear in members of $\mathbf{M_p}$, but not in members of $\mathbf{M_{pp}}$ are "uninterpreted" or "theoretical" concepts for the theory element **T**. They represent concepts that are "internal" to **T** in that they derive their meaning solely from their use in solving problems associated with **T** and, perhaps as well, from their role in theory elements that **T** interprets. This distinction recalls the distinction between "theoretical" and "observational" terms in traditional logical empiricism except that it is explicitly localized to a single theory element.

The intended applications **I** of a theory element are simply a sub-class of $\mathbf{M_{pp}}$. Intuitively, **I** is that sub-class of $\mathbf{M_{pp}}$ consisting of systems for which other theory elements can "in fact" (rather than just as a formal possibility) provide information. That is, members of **I** must ultimately be linked to "solved" problems in theory elements that interpret **T**. Other theory elements linked to a theory element **T** by interpreting links may be thought of as providing "data" about members of **I**.

Theory element $\mathbf{T} = \langle \mathbf{K}, \mathbf{I} \rangle$ is "about" **I** in the sense that it says something about it. The "claim" of theory element **T** is that all members of **I** can be filled out with the full (theoretical) vocabulary of $\mathbf{M_p}$ in a way that satisfies *both* laws **M** and constraints **C**. Very roughly, the problems for **T** are the members of **I** and solving such a problem is providing a specific configuration of theoretical vocabulary that fills out a member of **I** to a member of **M**. Generally, this problem solving occurs in a situation in which other such problems have already been solved so that the filling out is also required to preserve the satisfaction of **C** when it is added to the stock of already solved problems. More subtle accounts of scientific practice consider the possibility that problems are sometimes deleted from the stock of "solved" problems. We ignore these subtleties here.

The metaphor of problem solving as "filtering and transforming information" may be fleshed-out in the following ways. First, not all possible configurations of information transmitted by links *to* **T** may be amenable to "filling out" with theoretical concepts of **T** in the way just described. In most cases, the claim of **T** will be non-vacuous in that it will "reject" certain possible configurations of data. In this way, **T** acts as a filter. But, it also transforms the information it receives from its interpreting links in an essential way. Most obviously, the theoretical concepts that appear in successfully solved problems for **T** do not appear in any recognizable form in the data that sets the problems. More precisely, they can not be explicitly defined in terms used to describe this data. Yet it is these full-bodied solutions that **T** transmits to other theories for which it provides partial descriptions of data. Less obviously, the links themselves may serve to "transform" concepts from one theory into those of another.

We may now see, somewhat more clearly, how the intended applications $\mathbf{I}$ in $\mathbf{T} = \langle \mathbf{K}, \mathbf{I} \rangle$ are determined. $\mathbf{I}$ consists just of those members of $\mathbf{M_{PP}}$ that are (or can be) linked to *successful* intended applications $\mathbf{I}'$. In theories that are linked *to* (interpret) $\mathbf{T}$. Minimally, 'successful' means members of $\mathbf{I}'$ that have been filled out with theoretical concepts of $\mathbf{T}' = \langle \mathbf{K}', \mathbf{I}' \rangle$ in the appropriate way. Less minimally, successful applications of $\mathbf{T}'$ will have to be linked to successful applications of *other* theories (if any) besides $\mathbf{T}$ that $\mathbf{T}'$ may serve to interpret.

In this manner, the net surrounding a theory element $\mathbf{T} = \langle \mathbf{K}, \mathbf{I} \rangle$ may be seen as providing an intensional description of the intended applications $\mathbf{I}$. It does this recursively by telling us how to check a candidate for membership in $\mathbf{I}$ by tracing its connections through the links it has to other intended applications of other theory elements linked with $\mathbf{T}$. This tracing back terminates when it reaches theory elements $\mathbf{T}^{\sim}$ in which *extensional* descriptions of successful applications (members of $\mathbf{A}^{\sim}$) are found. Otherwise, at every theory element $\mathbf{T}$ in the recursive tracing back a new problem for $\mathbf{T}$ is generated.

ELEMENTARY BASIC RESEARCH PROGRAMS

Using this as the account of problems for a theory element just sketched, we now turn to the question of choosing the order of attacking these problems. Here too the sociology of science literature offers some guidance, though sketchy. A summary of this work ([12], p. 82) suggests that ". . . . two criteria were most frequently used in selecting from arrays of previously identified problems: (1) the assessed scientific importance of a problem . . . and (2) the feasibility of arriving at solutions". This summary suggests that the apparatus of microeconomic decision theory might be applicable to providing a more detailed theory about this kind of problem choice. It appears quite natural to identify "assessed scientific importance" as a value or utility and "feasibility of arriving at solutions" as a probability. For the moment we will leave open the question of just *whose* values and probabilities these might be and focus on the question of using the representation of scientific problem solving just sketched to provide a somewhat more precise formulation of this decision problem.

To begin, let us conceptualize problem choice as a resource allocation problem – one chooses to work on a problem by allocating resources to it. 'Resources' is to be understood intuitively to consist of personnel and equipment devoted to the solution of a problem. Let us also conceptualize the problem as one of choosing a "research program" – a plan for allocating resources to problems over some period of time.

We may consider one very simple kind of "research program" for a theory element $\mathbf{T} = \langle \mathbf{K}, \mathbf{I} \rangle$. We call these 'elementary research programs' to indicate that they are exclusively concerned with a single theory element. Elementary research programs are very likely an idealization in that no real piece of scientific activity may plausibly be modeled as an elementary research program. Nevertheless, they are a useful starting point for an analysis and may well prove to be fundamental parts of more complex research programs.

Let us suppose that the program begins with some problems for $\mathbf{T}$ already solved. That is, at time t there is a finite sub-class $\mathbf{A}_t$ of $\mathbf{I}$ for which there are at hand extensional descriptions of theoretical emendations of $\mathbf{A}_t$ that satisfy $\mathbf{M}$ and $\mathbf{C}$. The $\mathbf{A}_t$ could be the "paradigm" problem solutions which guide the practice of normal science on Kuhn's account, but they need not be. Generally, they just represent the stock of successful applications of $\mathbf{T}$ at any time t in the history of the community that practices science with theory element $\mathbf{T}$. We assume that $\mathbf{A}_t$ is non-empty and leave the limiting case of an empty $\mathbf{A}_t$ for special treatment. For simplicity, assume $\mathbf{I}$ is finite, $\mathbf{I} = \{i_n, \ldots, i_n\}$.

In the situation just described, a resource allocation problem for $\mathbf{T}$-problems may be formulated in this way. How is a finite amount of "resource" L to be distributed over the problems in $\mathbf{I} \sim \mathbf{A}_t$? At this point we need not be too specific about what L is. We need only to think of it as something measured by a monotone increasing real valued function. More subtle treatments might take it to be characterized by a real vector. For example, one might want to take time as a component of this vector requiring special treatment.

How should we conceptualize a "research program" aimed at solving problems in $\mathbf{I} \sim \mathbf{A}_t$? Most generally, one might conceive of a research program as a sequence of "allocation vectors":

(1) $r(1), r(2), \ldots, r(n), \ldots$

where each vector $r(t)$ represents an allocation of resources to members of **I** at stage t in the program. That is, $r(t)$ is of the form:

(2) $r(t) = \langle r(t, 1), r(t, 2), \ldots, r(t, n) \rangle$.

The resource vectors are subject to the following constraints.

(3) $\sum_{t,i} r(t, i) = L$.

(4) $r(t, i) = 0$, for all i in $\mathbf{A}_t$.

A somewhat more realistic formulation might replace '=' in (3) with '≤', but little insight appears to be gained by this added complexity. We remain vague about the interpretation of the "stages" in a research program.

If we let 'R' denote the class of all research programs satisfying the above conditions (1–4), then "research planning problem", at the scientific community level, may be conceived simply as choosing *one* among the feasible research programs in R. The one chosen should be among those that are optimal with respect to *some* value or "objective function" for the scientific community. Our purpose here is to describe this objective function.

We may proceed by introducing some notation to describe the "solution state" of problems in **I**. Let $Z = \{t_1, \ldots, t_m\}$ be the sequence of stages in the research program, $I = \langle i_1, \ldots i_n \rangle$ some arbitrary ordering of I and

(5) $s: Z \rightarrow \{0, 1\}^n$.

The value $s(t)$ is to be interpreted thus: problems corresponding to positions where 1's appear in $s(t)$ have been solved and those corresponding to positions where 0's appear have not. Each such function s describes a "possible problem solution history" or "solution sequence" and we may denote the set of all such histories by 'S'. Clearly, we idealize in regarding problems as simply "solved" or "unsolved".

This interpretation suggests that at least one additional condition might plausibly be imposed on members of S. The condition that "problems never become unsolved" may be expressed by:

(6) $s(t+1) \geqslant s(t)$,

where the partial ordering "$\geqslant$" on vectors means that their respective components are partially ordered by the usual $\geqslant$-relation on real numbers. This condition is only plausible on a rather strong interpretation of "solved". On this interpretation, a problem is simply not "solved" if there remains a possibility that a mistake was made which will be subsequently uncovered resulting in the problem's status reverting to "unsolved". With a weaker and more common interpretation of "solved", one would have to countenance the possibility that problems sometimes became "unsolved".

It should be noted that S will include solution sequences in which problems are solved "one-at-a-time". That is, $s(t+1)$ will contain exactly one more '1' than $s(t)$. But it will also contain solution sequences in which more than one problem becomes solved in a given time period. That is, $s(t+1)$ may contain several more '1''s than $s(t)$.

To begin formulating an objective function for an elementary research program, let us first suppose that some value *extrinsic* to theory element **T** is assigned to each problem solution vector which is represented by a function:

(7) $\quad U: \{0, 1\}^n \rightarrow \mathbb{R}.$

Values of different problems might be independent in the sense that $U(s(t))$ was always expressible as the sum of u-values for "elementary" solution vectors having only one non-0 component. We do not require this assumption. Intuitively, extrinsic values may be derived in some way from connections with other theory elements and/or technological considerations entirely extrinsic to "basic" **T**-science. We will return to consider their source below. Note that $U(s(t))$ represents the value of "having solutions to problems represented by $s(t)$" – not the value of some specific solutions. Note also that we consider the extrinsic values to be independent of the stage in the research program in which the problems are solved. $U(s(t_m))$ is the value of having solutions to the problems $s(t_m)$ that terminates the research program s. Thus it is plausible to define the value of s to be

(8) $\quad U(s) := U(s(t_m)).$

It might appear plausible to require further that the extrinsic value of having the solution to more problems is greater than that of having the solution to fewer. In view of (6), this would mean simply that,

$U(s(t+1)) \geqslant U(s(t))$. We avoid taking this explicitly Panglossian view of the value of basic science.

Turning to probabilities, suppose that the probability of solution sequence s, given resource allocation vector r is given by:

$$(9) \qquad \begin{aligned} P(s, r) = {} & P(s(t) \mid s(t-1), r(t)) \times \\ & P(s(t-1) \mid s(t-2), r(t-1)) \times \\ & \vdots \\ & P(s(1) \mid s(0), r(1)). \end{aligned}$$

That is, probabilities of problem solution histories are "Markov" in the sense that the probability of $s(t)$ depends only on $s(t-1)$ and the resource allocation vector at t. This is clearly an oversimplification in that it ignores investment in equipment at period t_i that may be used for solving problems in periods $t \geqslant t_i$.

Finally, suppose that, the probability of problem solutions $s(t)$, given the solution state $s(t-1)$ and the resource allocation vector $r(t)$:

$$(10) \qquad P(s(t) \mid s(t-1), r(t))$$

is a monotone non-decreasing, convex function of the resource allocation vector r, and nothing else. That is,

$$(11) \qquad P: S \times S \times R \rightarrow [0, 1]$$

is a conditional probability measure on S such that:

(12a) if $s(t) \leqslant s(t-1)$, then, for all $r(t)$,
$P(s(t) \mid s(t-1), r(t)) = 0$;

(12b) if $s(t) > s(t-1)$ and $r(t) > r'(t)$ then
$P(s(t) \mid s(t-1), r(t)) \geqslant P(s(t) \mid s(t-1), r'(t))$.

That is, no allocation of resources can result in solved problems becoming unsolved (cf. condition (6) above) and more is always more effective than less.

Under these assumptions, the expected value E of a resource allocation vector r is

$$(13) \qquad E(r) = \left(\sum_{s \in S} U(s) P(s, r) \right) - L$$

and we might plausibly take this expectation value to be the objective function for an elementary research program in the environment of exogenously supplied values $U(s)$. Recall that $U(s)$ is the value of the problems solved at the termination of the program r. To conceptualize "intrinsic scientific value" – termed just 'scientific value' in what follows, we might simply supppose that the exogenous values are uniform – i.e., $U(s) = U(S') = K$, for all $s \in S$. Then

$$(14) \qquad E(r) = K\left(\sum_{s \subset S} P(s, r)\right) - L$$

Intuitively, we might think of K as being the value that some "disinterested", external benefactor attaches to **T**-basic science – "disinterested" in the sense that she does not care which **T**-problems get solved.

Assuming that $\mathbf{r}^*$ is a research program that maximizes $E(r)$ in (14), we might then identify $\mathbf{r}^*(t, i)$ as the scientific value of having a solution to **T**-problem i at stage t. Intuitively, the scientific value of a problem in an elementary research program is just whatever resources would be allocated to its solution in an optimal research program. The content of this "definition" derives from the fact that we have a way of determining optimal research programs that is independent of knowing the scientific value of problems. On this account, the intrinsic scientific value of the solution to a problem is time dependent. I shall consider the intuitive plausibility of this below. In the case that there is more than one research program that maximizes $E(r)$, $r^*(i, t)$ might not be uniquely determined in this way. At this stage of the analysis, I simply note this possibility and proceed to ignore it.

It is convenient to extend this idea to define the **T**-relative, or local, scientific value of temporal stages in solution sequences. The obvious way to do this is simply to sum the value of $\mathbf{r}^*(i, t)$ over the problems with non-zero values in the vector $s(t)$. Thus, I define

$$(15) \qquad \mathbf{VT}_\mathbf{I}(x(t)) := s(t) \cdot r^*(t),$$

where '·' is the scalar product.

The optimization problem posed by (14) can be attacked by the usual microeconomic methods. However, an intuitive understanding is perhaps more important for our purposes. The basic intuitive idea in (14) is that not all solution sequences are equally likely. There is generally a "natural order" in the solution of problems. Solutions to some problems become easier (i.e., cheaper), once other problems have been solved. Optimal research programs are going to be those which address problems in this natural order.

This is neither a particularly novel nor profound idea, but, it does suggest a sharp, austere formulation of the concept of "scientific value" relative to a theory element **T**. Roughly, scientific value of unsolved problems is identified with the resources that would be allocated to their solution in some optimal research program in which external values are uniform. Intuitively, problems differ in scientific value because they differ in the degree to which they contribute to the solution to other problems. Still more intuitively, problems have scientific value just because their solution contributes to the solution of other problems. It is important to see that this concept of "relative" scientific value is independent of any assumption about the inherent social value of knowledge. Formally, this just means that it is independent of the value of **K** in (14). This independence is emphasized when we come to consider below how the probabilities in (14) might be determined solely by the formal properties of the theory element **T**.

With this intuitive understanding of our conception of scientific merit, it is not surprising that it turns out to be time dependent. When solutions to problems have only instrumental value as "means" to the solution of still other problems, it is natural to expect that this value will depend on which problems remain to be solved and, more specifically, which are next on the agenda of an optimal research program. One should expect that the scientific value of a problem is relative to the context of a specific research program *and* to a specific stage in that program. This suggests that optimizations in research planning should reassign priorities and reallocate resources frequently in the light of new information about the *actual* progress of problem solution. A more subtle analysis would countenance costs of information about the solution vector s as well as "reprogramming" costs.

The analysis of "scientific value" is relativized to a single theory element **T** simply because it is primarily with respect to single theory elements that our representation of scientific knowledge provides us with a characterization of scientific problems. It is also with respect to single theory elements that we have some clear idea of what might determine the probabilities of sequences of problem solutions. The most obvious idea is that these probabilities are determined by the constraints **C** that operate across sets of potential models. Intuitively, given some initial set of solved problems $\mathbf{A}_t$, the most probable solution sequences are going to be those that add new members to $\mathbf{A}_t$ that are linked to members of $\mathbf{A}_t$ by constraints. For example, consider a research program aimed at applying Daltonian stoichiometry to a class of chemical reactions larger than that to which it has already been successfully applied. (For details on this example see [1].) Common sense, historical data and formal analysis converge to suggest that the most economical research program begins with new reactions that involve at least some of the same compounds participating in the reactions to which the theory has already been successfully applied.

It is not obvious that constraints are the only factors relevant to probabilities of solution sequences for a single theory element $\mathbf{T} = \langle \mathbf{K}, \mathbf{I} \rangle$. The interpreting links that determine the intended applications I may operate in such a way that probabilities (as a function of cost) of acquiring data needed for problem solution differ among intended applications that are similarity situated with respect to the totally formal properties of T-constraints. For example, reliable pure samples of some, otherwise equally "interesting", compounds may differ widely in their cost. This suggests that our conception of an intertheoretical link might profitably be augmented with some kind of cost function.

Neither is it obvious how information about constraints, costs of interpreting links and perhaps other factors can be translated into probabilities of solution sequences. At best, these remarks suggest a way of posing the question.

BASIC RESEARCH PROGRAMS

Let us now consider how the concept of scientific value relative to an elementary research program for a theory element $\mathbf{T}$ may be extended to provide a more general concept of scientific value.

We may begin by generalizing our notion of an elementary research program. To do this, consider a situation in which a theory element $T = \langle \mathbf{K}, \mathbf{I} \rangle$ resides in a net of theory elements connected by interpreting links. Suppose that, in the course of carrying out a $\mathbf{T}$-elementary research program, additional data are required about member x of $\mathbf{I}$. We represent the procedure of acquiring these data in the following way. Data about x are provided by potential model x' – linked to x – in some theory element $\mathbf{T}'$ that interprets $\mathbf{T}$. Potential model x' may be among the solved problems of $\mathbf{T}' - \mathbf{A}'_t$. If so, the required data are just "exported" to $\mathbf{T}$. If not, the x' is put on the "local research agenda" for $\mathbf{T}'$. More precisely, the $\mathbf{T}'$-scientific value of the problem x' has been augmented by virtue of its being called for as a part of the $\mathbf{T}$ local research program. Solving problem x' may require solving problems in other theories besides $\mathbf{T}$ that $\mathbf{T}$ interprets and, as well, solving problems in theories that interpret $\mathbf{T}'$. In this way, local research programs in $\mathbf{T}$ may be viewed as recursively generating local research programs in other theories linked with $\mathbf{T}$. That is, global research programs can be viewed as resulting from aggregation of recursive propagation of local research programs. These ideas are made somewhat more precise in [9].

Given this picture of global research programs, an obviously attractive way to generalize our notion of scientific value is to envision that it propagates in a recursive and cumulative way through the net of theory elements. That is, the $\mathbf{T}$-scientific value of the members of $\mathbf{I}$ in $\mathbf{T} = \langle \mathbf{K}, \mathbf{I} \rangle$ call it $U[\mathbf{T}]$ gets added to the $\mathbf{T}'$-value of the members of $\mathbf{I}'$ in $\mathbf{T}' = \langle \mathbf{K}', \mathbf{I}' \rangle$ when $\mathbf{T}'$ interprets $\mathbf{T}$. Intuitively, members of $\mathbf{I}'$ are valuable as parts of elementary research programs relative to $\mathbf{I}'$ and valuable as well because they are linked with elementary research programs relative to $\mathbf{T}$. In this way, members of $\mathbf{I}'$ would accumulate scientific value from all the theories that $\mathbf{T}'$ interpreted and, in turn, pass the *sum* of these values on to the intended applications in the theory elements that interpret $\mathbf{T}'$.

Let us now consider the global scientific value – $\mathbf{V_g}$ – of a solution vector for $\mathbf{T}$. The intuitive ideas just outlined suggest that $\mathbf{V_g}$ should satisfy the following condition:

$$\mathbf{VT_g}(s(t)) = \mathbf{VT_1}(s(t)) + \sum_{\mathbf{link}(s(t),\, s'(t))} \mathbf{VT}'_\mathbf{g}(s'(t)). \tag{18}$$

The notation '$\mathbf{link}(s(t), s'(t))$' means that $s'(t)$ is some solution vector for a theory $\mathbf{T}'$ that $\mathbf{T}$ interprets and all non-zero components of $s'(t)$ are linked to non-zero components of $s(t)$.

Note that it is the global scientific value of the interpreted theories $\mathbf{T}'$ that appears on the right side of (18). Intuitively, this means that global scientific merit for $\mathbf{T}$ is accumulated from all the theory elements to which $\mathbf{T}$ is "forward connected" by interpreting links. This view has the intuitive advantage of making scientific value of a given problem a function of the "density" of the links between this problem and other problems. (See [10].)

Roughly, this suggests that one begins calculating global scientific merit with those theories that do not interpret any other theories (if such exist). For these theories global and local scientific merit are identical. One then goes one level lower to the theories that interpret these and solves a maximization problem isomorphic to (13) with $U(s)$ replaced by the $\mathbf{VT}_g(s)$ given by (18). Then one iterates the same procedure at the next lower level until reaching the theory element of interest.

Aside from depending on heroic assumptions about our knowledge of the global structure of science and the implications of local structure (e.g., constraints) for solution probabilities, this account of global scientific value suffers from an additional defect. It depends on the assumption that interpretation links do not form closed loops in that procedures for evaluating the sum in (18) will only terminate under this assumption. It is far from obvious that this is true. (See [1], Chap. 8.) It would be nice to have some kind of measure of global scientific value that did not depend on this assumption. Intuitively, what's wrong with (18) is that theory elements "far away" from **T** contribute just as much as those "close to" **T** to the scientific value of problems in **T**. Adding some kind of "attenuating factor" like 1/(number of links in chain connecting **T** and **T**′) might be more plausible. Still another approach would be to place some necessary conditions on global scientific value that related it to local scientific value via something like difference equations without necessarily determining its value. Intuitive considerations make this attractive, but do not appear to indicate an obvious way to proceed.

SPECIAL LAWS AND LAW DISCOVERY IN BASIC RESEARCH PROGRAMS

Up to this point, our discussion has focused exclusively on the solution of problems in an environment in which a single law is employed. Most scientific practice is not so simple. Theories with conceptual sophistication and complexity – for example classical equilibrium thermodynamics – typically consist of several theory elements each representing different laws linked by a "specialization link" into a tree like graph which coexists with the net of interpretation links. Specialization linked theory elements constitute a "theory element family" in that they all have the same conceptual apparatus $\mathbf{M_p}$ and (non-theoretical conceptual apparatus $\mathbf{M_{pp}}$. They differ only in their laws **M** and constraints **C**.

This specialization structure plays a role in the problem solving process in the single theory elements. Roughly, specialization provides additional possibilities for determining the values of theoretical concepts. These values may be passed through the specialization from one theory element net to facilitate the solution of problems in other theory elements. For example, masses discovered in Hooke's law systems (spring balances) may be employed in other Newtonian systems characterized by other force laws.

The account of basic research programs offered above needs to be emended to include this kind of activity. But, more importantly, this possibility suggests that a more radically different kind of scientific activity needs to be considered – the activity of discovering new theory elements related to existing ones by specialization links. On the account of "scientific problems" suggested above, discovering new scientific laws couched in a given conceptual framework is just a different kind of activity than discovering how to apply a known law to a new situation.

How can we appraise the scientific value of this "law searching" kind of activity? One approach is to regard it as "overhead" on the kind of problem solving activity we have already considered. The clue here is that having special laws always (in principle) or usually (in practice) makes the solution of these "first order" problems easier (cheaper). Formally, this results from their being additional possibilities of determining the values of theoretical concepts and transferring this information to where it can be used in (first order) problem solution.

Can this insight be translated, even in principle, into a quantitative measure of the value of having a new special law? For any specific additional special law, it is not implausible to suppose that one might quantify its contribution to (first order) problem solving. At least, this would be no more difficult than for constraints. However, the search for special laws is intuitively not an attempt to *justify* the addition of a specific, already formulated law. Rather, it is the search for *some* new special laws of unspecified form. How can we calculate the scientific value of such a search?

NOTE

* The research reported here was partially supported by a grant from the EXXON Foundation.

REFERENCES

[1] Balzer, W., C.-U. Moulines and J. D. Sneed: 1987, *An Architectonic for Science*. Reidel, Dordrecht.

[2] Gieryn, Thomas F.: 1978, 'Problem Retention and Problem Change in Science', *Sociological Inquiry* **48**, 96–115.

[3] Hobbs, J. R. and R. C. Moore: 1985, *Formal Theories of the Common Sense World*. Ablex, Norwood NJ.

[4] Krantz, D. L.: 1965, 'Research Activity in 'Normal' and 'Anomalous' Areas', *Journal of the Behavioral Sciences* **1**, 39–42.

[5] Kuhn, Thomas, H.: 1962, *The Structure of Scientific Revolutions*, University of Chicago Press, Chicago.

[6] Martin, B. R. and J. Irvine: 1981, 'Internal Criteria for Scientific Choice: An Evaluation of Research in High-Energy Physics Using Electron Accelerators', *Minerva* **19**, 408–32.

[7] Merton, R. K.: 1957, 'Priorities in Scientific Discovery: A Chapter in the Sociology of Science', *American Sociological Review* **22**, 635–59.

[8] Mulkay, M. J., G. N. Gilbert and S. Woolgar: 1975, 'Problem Areas and Research Networks in Science', in *Sociology*, *The Journal of the British Sociological Association*, **9**, 187–203.

[9] Sneed, J. D.: 1986, 'Machine Models of the Growth of Knowledge: Theory Nets in PROLOG' to appear in *Criticism and the Growth of Knowledge*: *20 Years After*: *Proceedings of an International Conference in Memory of Imre Lakatos*, Thessaloniki, Greece.

[10] Weinberg, Alvin M.: 1984, 'Values in Science: Unity as a Criterion of Scientific Choice', *Minerva* **22**, 1–12.

[11] Weinstein, Deena: 1976, 'Determinants of Problem Choice in Scientific Research', *Sociological Symposium* **16**, 13–23.

[12] Zuckerman, H.: 1978, 'Theory Choice and Problem Choice in Science', *Sociological Inquiry* **48**, 65–95.

Manuscript received 25 January 1988

Colorado School of Mines
Department of Humanities
Golden, Colorado-80401
U.S.A.

LESZEK NOWAK

ON THE (IDEALIZATIONAL) STRUCTURE OF ECONOMIC THEORIES*

The present paper is devoted to an analysis of the internal structure of theories built in economics. Of different methodological topics connected with economic theories only the structural ones are discussed here.[1] I begin with a rough outline of different deformational procedures in order to show the one, viz. idealization, that seems to be a proper starting point for a more detailed analysis. Then I shall try to reconstruct the structure of economic theories in what seems o be its main features (Section II) and variants (Section III).

I. SOME EPISTEMOLOGICAL ASSUMPTIONS

The Family of Deformation Procedures

Let us consider the set U of properties (magnitudes, factors) and an object, a, qualified as an existing one, with a certain space of properties occurring on it in some degree. It might be said that potentialization of a consists in postulating a possible object, a', which differs from a in that a' has some property P in a degree different from that in which a has P; if the new degree, that pertaining to a', is greater (resp. lesser) than that pertaining to a, we shall speak of positive (resp. negative) potentialization.

The extreme case of negative potentialization might be called abstraction.[2] Under abstraction, a given object is counterfactually postulated to have one of its properties in the minimal ("zero") degree. In other words, an abstract of a given object is such a possibile of the latter which differs from it not in the space of properties but in the minimization of some of them. On the other hand, mythologization is the extreme case of the positive potentialization. It might be identified with maximization of a magnitude, so that a given object is counterfactually required to possess at least one of its parameters in the highest degree. A mythical type of a given object possesses then the same space of properties as the latter, but under some respects it is

Erkenntnis **30** (1989) 225–246.

extremely "exaggerated" in comparison to the actual state of that object.

Potentialization, either negative (with abstraction) or positive (with mythologization), is a soft-deformational procedure as it preserves the body of properties an initial being is supposed to possess; the only changes it postulates concern the degree in which given properties are to pertain to the considered objects. Meanwhile, the hard-deformational procedures consist actually in denying the given equipment of features of the object under consideration.

And so, reduction consists in that a given object is counterfactually postulated not to have some of the properties it actually has. That is, it is counterfactually assumed that its space of properties diminishes under such and such respect. In other words, a reduct a' of a is supposed to be a possible object lacking some properties the object a has; if a has n properties, there are reducts of the first level with $n-1$ properties, of the second one with the spaces composed of $n-2$ features, etc. The reduct of the nth level, that is, equipped with none of the properties of the universe of features U, deserves the name of the nothingness (in U). Notice that abstraction concerning property P does not neglect it, as it ascribes to a given object the property P in the minimal ("zero") degree, while reduction deprives this object of property P. In the former case the influence of P upon other properties is omitted whereas in the latter one the very existence of P (on the given object) is denied.

A counterpart of reduction might be called transcendentalization. Under transcendentalization, a given object is counterfactually postulated to have properties that it does not actually possess at all. A transcendentale of object a is then an object a' of the same logical type equipped with a larger space of properties than the object itself. Similarly as in the case of abstracts, transcendentalia (relative to the initial object) are of increasing levels. That of the highest level, i.e., possessing all the features constituting the given universe U of properties deserves the name of the absolute (in U).

A Remark on the Nature of Scientific Cognition

Let us comment a bit on the above distinctions. The four elementary deformational procedures – negative and positive potentializations, reduction and transcendentalization joined on par in compound ones seem to be characteristic for main domains of spiritual activity of man. One might even conjecture that a necessary condition for an activity to belong to (the spiritual) culture is to employ some of the deformational procedures.

For instance, application of fictionalization, that is, jointly reduction and positive potentialization (in particular, mythologization) seems to be characteristic for arts. A good caricature presupposes evidently reduction – some of the properties of the portrayed person are omitted, and mythologization – the remaining ones are exaggerated. When a writer collects the life stories of many persons in order to find in them elements – appropriately "coloured" – of the life-story of his protagonist, his mental activity might be likely reconstructed as fictionalization of the recorded material.

The domain of faith in turn, seems to involve absolutization, that is, the joint use of transcendentalization and positive potentialization, including perhaps mythologization. This appears in the domain of religion – the creatures postulated there are ontologically richer than the actual beings and their properties are to be of the "highest"-type: the Highest Power, the Highest Wisdom, the Highest Goodness, etc. Also the layman's faith presupposes this – Lukacs Proletariat which possesses its limit-true-consciousness and is able to break the laws of history creating new ones is nothing but absolutization of the actual working class. The faith appears so long as the inclination of a person to apply absolutization in the way prescribed by the faith is lasting.

In contradistinction to the domains of arts and faith, the peculiarity of cognition seems to be an application of idealization, that is, both reduction and negative potentialization (in particular – abstraction). This applies even to the level of common sense cognition – stereotype is nothing but a reduct minimizing additionally untypical features of what is stereotypized – but its proper field of application is evidently science. For instance, the mass point, is both a reduct and an abstract of physical bodies – it possesses only some physical properties (other properties are reduced, such as chemical, possibly biological or psychological etc. properties), but even some of those which can be meaningfully ascribed to it, e.g., space dimensions, are postulated to appear in the zero degree. Mass point, and all similar scientific constructs, are then ideal types of the actual objects. Let us look at the procedure of idealization in a bit more detailed manner.

On the Notion of an Essential Property

First thing that must be decided is why making idealizations we reduct some properties whereas some others are only abstracted from (i.e., equalized to zero). This leads to the notion of essential property. The notion might be, and is, explicated in different ways (cf. presentation in Loux 1983, p. 16ff). I would like to consider here one of possible ways of explication not declaring, however, any kind of "faith" in it. Conversely, what is insisted in this connection is that none of the methodological hypotheses that will be put forward below depends on the contents of the discussed explication.

Let it be given the set of properties (magnitudes, factors) U and a property F from U. A factor G will be said to be essential for F iff for every value v of G there exists a non-empty subset of values V of F such that for every object x, if $G(x) = v$, then it is not the case that $F(x)$ belongs to V; the set V is termed the range of exclusion of G relative to F. In other words, if one factor G influences upon another F, then it is excluded that given a definite value of G, F can take an arbitrary value. Conversely, if G were non-essential for F, then for some values of G the magnitude F could take quite an arbitrary value: that is, V would be an empty set. It is obvious that the range of exclusion V has its maximum – it can be at most equal to the set of values of F minus one-element set of values $\{w\}$, w being the one value F in fact assumes. In this case we have to do with the strict determination: F would be then univoqually determined by G; that is, for every x, if $G(x) = v$, then $F(x) = w$. Singular factors influencing a given one have their ranges of exclusion larger than the minimum (the empty set) and not exceeding the maximum (universe of values minus one); normally, the maximum for singular factors is not achieved at all.

Let us consider the magnitude F and the set of all the factors essential for F. The assumption of determinism says that for every magnitude F there exists a set of magnitudes essential for F of such a kind that their joint influence upon F is maximal, that is, if all the factors take on their values on the given object, then the object has one value of F. Such a set will be termed the space p_F of essential factors for F.

The cognitive sense of reduction becomes hopefully a little more clear now. Its aim is to divide, given the property F, the set U of properties upon those that are essential for F and the remaining ones. The latter are to be reduced. In this manner a possibile is postulated (the reduct of the initial, actual object) that is claimed to possess all and only those properties that are essential for F. It is evident that a researcher might be wrong in making such a division, that is, that he reduces some properties which are in fact essential for F or that he includes into the space P_F those whose influence upon F is none.

On the Essential Structure of a Factor

The question arises what is the cognitive sense of the procedure of abstraction (i.e., the counter-factual zero-ing some magnitudes). It seems that the answer must refer to hierarchization of the space of essential properties (for a given one). Let us consider the following tentative explication.

A factor G is said to be more essential for F than a factor G' iff, for every values v and v' of G and G' correspondingly, the range V' of exclusion of G' (relative to F and v') is a proper sub-set of the range V of exclusion of G (relative to F and v). In other words, the more significant for F a given factor is, the greater is the set of values of F excluded by the fact that this factor takes on a given value. Obviously, it might be so that in the space of essential factors for F there are some magnitudes none of which is more essential than the other for F; they will be labelled as equi-essential for F. Now, the hierarchy of sets $E_1, E_2, \ldots, E_n$ in which every set contains factors equi-essential for F but those of E_i are more essential for F than those of E_{i-1} is termed the essential structure S_F for F. The principal factors are those which are more essential than all the rest, i.e., those of E_1, the rest being secondary ones – relative to F.

It is obvious that nobody knows a priori which factors are in fact principal for a given one; everything is a matter of "trial and error".[3] What is important here is, however, that the sense of the procedure of idealization becomes, I conjecture, more easy to understand: employing both reduction (of factors identified as unessential) and abstraction (of factors identified as secondary) it aims at a reconstruction of the essential structure for the investigated magnitude. Yet, everything it might bring is an image of that structure, adequate or not.

The Structure of Scientific Explanation

Having adopted the above assumptions, the structure of scientific explanation may be presented as follows.

(I) Let it be so that a researcher wants to explain the phenomenon F in the class of objects R. On the foundation of the background knowledge some factors say, H, p_k, $p_{k-1}, \ldots, p_2, p_1$ are distinguished as essential for F (an image $I(P_F)$ of the space of essential factors for F). Obviously, $I(P_F)$ may remain to the P_F in all possible relations including identity on the one side and exclusion on the other.

(II) On the same foundation, the image of the space of essential factors for F is ordered as to the strength of influence upon F and an image $I(S_F)$ of the essential structure for F is hypothetically proposed.

(III) All the factors which are not considered to be essential for F (i.e., those which do not belong to $I(P_F)$) are reduced. All the factors considered to be secondary for F are abstracted from by adopting conditions of the form:

(id) $\quad p_i(x) = 0$

postulating counterfactually that an essential property p_i for F does not influence F are termed – somewhat unfortunately in the light of the intuitions outlined above – idealizing conditions. Then a simple dependence between F and what is believed to be its principal factor is being hypothetically proposed. In this manner an idealizational law (like statement):

($\mathbf{T}_k$) $\quad$ if $R(x)$ and $(p_1(x) = 0$ and $\ldots$ and $p_{k-1}(x) = 0$ and $p_k(x) = 0)$, then $F(x) = f_k(H(x))$

is put forward.

(IV) The law ($\mathbf{T}_k$) is then concretized with respect to the influence of less essential factors. And so, the condition $p_k(x) = 0$ is removed and an appropriate correction to the consequent of ($\mathbf{T}_k$) is being introduced; hence the first concretization of law ($\mathbf{T}_k$) is of the form:

($\mathbf{T}_{k-1}$) $\quad$ if $R(x)$ and $p_1(x) = 0$ and $\ldots$ and $p_{k-1}(x) = 0$ and $p_k(x) \neq 0$, then $F(x) = f_{k-1}(H(x), p_k(x))$

etc. till the final concretization which is a factual statement:

($\mathbf{T}_0$) $\quad$ if $R(x)$ and $p_1(x) \neq 0$ and $\ldots$ and $p_k(x) \neq 0$, then $F(x) = f_k(H(x), p_k(x), p_{k-1}(x), \ldots, p_1(x))$.

The statements $(\mathbf{T}_k)$, $(\mathbf{T}_{k-1}), \ldots, (\mathbf{T}_0)$, form an idealizational sequence.[4] It constitutes a skeleton of explanation of phenomena of the F-type (F-phenomena), the latter being understood as facts of pertaining of F to the investigated objects. If the known F-phenomena are explained by such a sequence, it becomes confirmed. If there appear some which cannot be explained in this manner, the sequence becomes – if some additional requirements are met (cf. (1980), Chaps. 6 and 10) – disconfirmed.[5] If the latter is the case, this testifies to the fact that either the forms of dependencies have been chosen wrongly or the initial reduction has been proceeded too early leaving aside factors essential for F or counting as essential for F those which are not.

There is no need to add that in the actual research practice the stages of the explanatory procedure are passed through automatically. It is a matter of a philosopher of science to reveal all of them or, rather, all of those which are capable of becoming conceptualized in the light of his epistemological assumptions. It is quite trivial that the latter might be false and that his decision to concretize the "methodological meta-idealizations" he puts forth might be too early.

II. THE BASIC STRUCTURE OF AN ECONOMIC THEORY

Idealization and Economic Modelling

The scheme of explanation outlined above seems to agree with the main intention of the model-building in economics as presented by outstanding economists. For example, according to E. D. Domar: constructing of a model, or any theory, consists in abstracting from an enormous and complicated mass of facts – termed the reality – several simple, easy to manage key matters which become for certain purposes a substitute of the reality itself. The simplification is the very heart of the process. And deciding of how and where to simplify and which variables are to be considered and which are to be eliminated – which is the very essence of the theoretical work – was and will remain a really subtle skill (Domar 1962, p. 60). One might also add that, at least in the realm of sciences of man, the economy seems to possess a historical priority in applying the method of idealization as it had been introduced in the theoretical work of K. Marx under the name of the "method of abstraction" (cf. a reconstruction in 1970, 1971a–b). All this suggests to hypothetically assuming that idealization is the "very essence" of the economic modelling.

The Adopted (Methodological) Simplifications

It is obvious that idealization is not everything that a theoretician of the economy does. Yet, in the same way as he is inclined to abstract from all the influences he decides to be secondary, so should the methodologist try to conceptualize his research practice. Therefore, we neglect in the beginning all the procedures except idealization. In order to do that our theoretician has to be very strongly simplified.

And so, we adopt a simplification that the background knowledge of our ideal economist (i.e., one with which he begins to construct explanation of F-facts) does not contain any other theory (the simplification **A**). It is assumed, then, that he works on the "theoretical fallow" having at his disposal only his philosophical presumptions and observational knowledge of singular facts.

Moreover, it is assumed that our researcher always can determine – truely or falsely – the way in which factors considered by him as secondary for a given magnitude influence it (simplification **B**). In other words, our researcher might be mistaken as to the assessment of the way a given factor influences the magnitude F but it is postulated that he has a definite view about that.

According to the next simplification, the researcher can always count factors that seem to him to be secondary (assumption **C**). Obviously, also this condition is an unrealistic one as in the standard theoretical situation there are very many "disturbances" the researcher cannot identify.

The next simplification claims that the only goal of the economist is to explain phenomena (assumption **D**).

It is obvious that much more simplifications are silently adopted here.[6]

The Deep Form of an Economic Theory: Model-I

Let us begin with the most idealized form of the economic model, namely that revealing exclusively the working of idealization as the only theoretical procedure.

Usually, not single factors but some wholes of them are investigated, and not single statements but some wholes composed of them are, hypothetically, proposed in the theory of economics. A criterion for characterizing such wholes as objects of modelling in science seems to be the following. Such factors are looked for that are principal ones for one another, and therefore modelling such a set of factors need not go beyond the modelled set of them; this is the case at least at the most idealized level of analysis. Let us conceptualize these intuitions a bit.

A set S of factors is said to be a system iff, for every F of S, the set of all factors principal for F is a subset of S. Correspondingly, an image of the system is termed a set $I(S)$ of factors such that, for every F of $I(S)$, the set of factors considered to be principal ones for F is included in $I(S)$. A theoretical system over $I(S)$ is a sequence of sets of statements $\mathbf{Q}_1, \mathbf{Q}_2, \ldots, \mathbf{Q}_n$, which meets the following conditions:

(a) for every magnitude of a statement of $\mathbf{Q}_1$, all the independent magnitudes shown in it belong to the set $I(S)$; the statements of $\mathbf{Q}_1$ are idealizational laws (i.e., are equipped with the maximal amount of idealizing conditions of all the statements of the system);

(b) the set $\mathbf{Q}_{i-1}$ is composed of statements which are concretizations of those of $\mathbf{Q}_i$ ($i = 1, \ldots, n$).

The deep form of the economic models (model-I) can be identified with a theoretical system in the outlined sense. Its aim is to find an objective system of factors and to reconstruct interdependences between them.

Let us add that the sense in which model-I is the deep, or basic, form of the economic models cannot be comprehended in terms of its typicality. Quite the contrary, it is rather doubtful whether the simplified forms defined by only one relation, that of concretization, appear in the actual practice of economic model-building at all. What is meant by saying this is that model-I is the simplest methodological scheme from which more realistic models, that are therefore more often met in the actual research practice are obtainable by removing the adopted simplifications. In other words, it is supposed that if simplifications (**A**–**D**) were satisfied all theories would be models-I, that is, they would possess their deep form.

The Theoretical Role of Deduction (*Model-II*)

Let us remove simplification (**A**). As long as it was in force, our idealized theoretician was applying the simple hypotheticist rule "invent hypotheses and test them against empirical data" (Popper 1959). After removal of (**A**) the situation changes: the research may use accumulated theories of his (enlarged) background knowledge. Obviously, the "accumulated theories" denote those which are admitted on the strength of our previous considerations, that is, models-I. On the present stage of abstraction our now (less) ideal theoretician applies another rule of theory-building: "deduce a model-I of a given system from the body of accumulated models-I; only if the latter turns out to be too weak for that, invent hypotheses and test them against empirical data". The theoretical role of deduction consists then in enlarging the body of the accumulated models-I.

The procedure of the theory-building on the present stage of abstraction is then the following.

(1a) Of the accumulated knowledge a set of idealizational statements is selected, all being in force under the same idealizing conditions; at most some of the latter might be irrelevant for some of them, but for every such a condition there exists a statement for which it is relevant. (1b) From the body of statements conclusions concerning the magnitudes the theoretician is interested in are derived.

(2a) Those assumptions are removed step by step and appropriate corrections are introduced, that is, the idealizational statements are concretized; at most those steps in concretization that involve the removal of irrelevant idealizing conditions turn out to be degenerate concretizations (i.e., possessing the same formula in the consequent as the concretized statement of the higher idealized order). (2b) From so obtained concretizations the consequences concerning the same magnitudes are deduced; they are valid on the appropriate lower level of abstraction.

(3) This procedure is repeated until the level of factual statements is attained.

According to this model-II will be identified with the following structure of statements:

$$(\mathbf{Q}_k, \mathbf{C}_k)$$
$$(\mathbf{Q}_{k-1}, \mathbf{C}_{k-1})$$
$$\dots\dots\dots\dots$$
$$(\mathbf{Q}_0, \mathbf{C}_0)$$

in which

(a) $(\mathbf{Q}_i, \mathbf{C}_i)$ are pairs of sets of idealizational statements $(k \leqslant i \leqslant 1)$; $(\mathrm{Q}_0, \mathbf{C}_0)$ is a pair of sets of factual statements;

(b) statements of $\mathbf{Q}_j$ (assumptions of the jth chain) are independent one from another whereas those of $\mathbf{C}_j$ (solutions of the jth chain) are consequences of the former;

(c) each statement of $\mathbf{Q}_{j+1}$ is a (strict or degenerate) concretization of some statement of Q_j $(k < j \leqslant 0)$.

Models-II are more realistic methodological constructs than models-I because they involve not only the relation of concretization but also that of entailment, which plays such an important role in the structure of economic theories.

Approximation (*Model-III*)

In spite of this, the structure of models-II is still highly abstract and very far from the actual complication of the economic theories. Let us then make it more realistic removing simplification (B). At the present stage of abstraction our theoretician has information – true or not – as to the working of only some factors he considers to be secondary for the investigated ones.

Let it be so that the researcher's background knowledge contains such information as to some most significant factors for F, say p_k, $p_{k-1}, \ldots, p_{i+1}$, and lacks any informations of the kind concerning the remaining magnitudes essential for F, viz. $p_i, \ldots, p_2, p_1$. Then he is unable to concretize the hypothesis:

($\mathbf{T}_i$) if $R(x)$ and $p_1(x) = 0$ and ... and $p_i(x) = 0$ and $p_{i+1}(x) \neq 0$ and ... and $p_k(x) \neq 0$, then $F(x) = f_i(H(x), p_k(x), \ldots, p_{i+1}(x))$.

Nonetheless, ($\mathbf{T}_i$) might be approximated, that is, referred to those facts in which the effects of working factors $p_1, \ldots, p_i$ are "small enough" so that one might expect the deviations between the left and right side of the formula of the consequent of ($\mathbf{T}_i$) do not exceed the "sufficiently small limits". In other words, the approximation of ($\mathbf{T}_i$) is of the form:

($\mathbf{AT}_i$) if $R(x)$ and $p_1(x) \leqslant a_1$ and ... and $p_i(x) \leqslant a_i$ and $p_{i+1}(x) \neq 0$ and ... and $p_k(x) \neq 0$, then $F(x) \approx f_i(H(x), p_k(x), \ldots, p_{i+1}(x))$

in which $a_1, \ldots, a_i$ are constants. The sequence of statements $(\mathbf{T}_k)$, $(\mathbf{T}_{k-1}), \ldots, (\mathbf{T}_i)$, $(\mathbf{AT}_i)$ is termed an approximation sequence.

Approximation[7] is quite a typical procedure applied by the economist because of the necessary "unrealistic" nature of the assumptions of economic theories. M. Friedman writes:

> the relevant question to ask about the 'assumptions' of a theory is not whether they are descriptively 'realistic', for they never are, but whether they are sufficiently good approximations for the purposes in hand. And this question can be answered only by seeing whether the theory works, which means whether it yields sufficiently accurate predictions. (Friedman, 1953, p. 15)

One might take a risk claiming that the intention of statements of the kind, so often in the writings of economists, can be interpreted in terms proposed above. This enables us to propose the following definition. A model-III will be termed the sequence:

$$\begin{array}{l} (\mathbf{Q}_k, \mathbf{C}_k) \\ (\mathbf{Q}_{k-1}, \mathbf{C}_{k-1}) \\ \cdots\cdots\cdots\cdots \\ (\mathbf{Q}_i, \mathbf{C}_i) \\ (\mathbf{AQ}_i, \mathbf{AC}_i) \end{array}$$

which differs from model-II only in that the statements of the set AQ_i (resp. AC_i) are approximations of those of Q_i (resp. C_i). That is, concretization works till the i-th level of abstraction and then is replaced with approximation.

And this is what seems to agree with intuitions the outstanding economists refer to. For instance, T. Koopman states:

> we look upon economic theory as a sequence of conceptional models that seek to express in simplified form different aspects of an always more complicated reality. At first these aspects are formalized as much as feasible in isolation, then in combinations of increasing realism. Each model is defined by a set of postulates, of which the implications are developed to the extent deemed worthwhile in relation to the aspects of reality expressed by the postulates. The study of simpler models is protected from the reproach of unreality by the consideration that these models may be prototypes of more realistic, but also more complicated, subsequent models. The card file of successfully completed pieces of reasoning represented by these models can then be looked upon as the logical core of economics, as the depositiory of available economic theory. (Koopmans 1957, p. 142/143)

What is said here is that the structure of economic theory is composed of the following three relations: (a) concretization – "models" (in our terms, chains of them) are put in a "sequence of increasing realism", (b) entailment – deducibility within singular "models" (chains) and (c) approximation – none of the "models" (chains) grasps the whole complexity of the actual empirical domain which is "always more complicated" and therefore can at most approximate it.

And it appears that model-III itself gives a sufficient approximation to the general structure of economic theories, that is, the structure common to all economic theories. In this sense model-III is supposed to represent the basic structure of the economic modelling.

Obviously, model-III is far from reproducing all the methodologically significant structural properties pertaining even to the typical economic theories. Yet, if I am not mistaken, it appears to be a general pattern from which different types of economic theories are obtainable through taking into consideration some additional structural dimensions.

III. SOME TYPES OF ECONOMIC THEORIES

Deterministic and Probabilistic Theories

Let us remove simplification (**C**) adopting realistically that in some cases the investigator is unable to identify all the secondary factors influencing a given phenomenon. Our investigator knows thus that there are some factors besides those shown in his image of the essential structure (as principal or secondary) that are essential for a given magnitude. Yet, he is unable to enumerate them, not to mention determining their influence upon the investigated magnitude. Factors of this kind will be termed interfering factors for the magnitude in question. The set i_F of interfering factors for F can be identified as the difference between the space P_F of essential factors for F and its image $I(P_F)$ established by our researcher.

The expression

$$E - i(x)_F = 0,$$

will be read: every factor of the set i_F takes on zero value for object x. If the researcher adopts this condition being convinced that the set i_F is non-empty, the expression will be called a semi-idealizing condition. The general statement containing in its antecedent a semi-idealizing condition will be called a semi-idealization statement. If it contains additionally some idealizing conditions, the statement is termed a semi-idealizational statement of the first kind. If this is not the case, the statement in question is a semi-idealizational statement of the second kind.

Let us consider the following semi-idealization statements:

(**ST**$_k$) if $R(x)$ and $E - i(x)_F = 0$ and $p_1(x) = 0$ and ... and $p_k(x) = 0$, then $F(x) = f_k(H(x))$

....

(**ST**$_i$) if $R(x)$ and $E - i(x)_F = 0$ and $p_1(x) = 0$ and ... and $p_i(x) = 0$ and $p_{i+1}(x) \neq 0$ and ... and $p_k(x) \neq 0$, then $F(x) = f_i(H(x), p_k(x), \ldots, p_{i+1}(x))$

(**SAT**$_i$) if $R(x)$ and $E - i(x)_F = 0$ and $p_1(x) \leqslant a_1$ and ... and $p_i(x) \leqslant a_i$ and $p_{i+1}(x) \neq 0$ and ... and $p_k(x) \neq 0$, then $F(x) \approx f_i(H(x), p_k(x), \ldots, p_{i+1}(x))$.

(**ST**$_k$), ..., (**ST**$_i$) are semi-idealizational statements of the first kind whereas (**SAT**$_i$) is that of the second kind: it takes into account the influence of the principal factor and the influence of the secondary ones (some in the strict, the rest in the approximate manner) but it does not account for any interferences. The univocal dependence f_i which is claimed to hold in (**SAT**$_i$) is subject to an influence of interfering factors of the set i_F. When the total influences of the factors i_F cancel each other, the dependence f_i holds but when some interferences prevail the remaining ones, the dependence f_i might not appear. The less essential factors belong to i_F, the higher is the percentage of the cases of the first type where the dependence f_i occurs. If more essential factors belong to i_F, then the percentage of occurrings of f_i is smaller. The ratio of the occurrings of f_i to the total number of elements of the universe of discourse (i.e., objects satisfying the realistic condition $R(x)$) is a relative frequency of the realization of f_i. Hence, accounting for the influence of interfering factors leads to a probabilistic statement:

(**PAT**)$_i$ the probability of

$$F(x) \approx f_i(H(x), p_k(x), \ldots, p_{i+1}))$$

on the condition that

$$R(x) \text{ and } S - i(x)_F \neq 0 \text{ and } p_1(x) \leqslant a_1 \text{ and} \ldots \text{and } p_i(x) \leqslant a_i$$

and $p_{i+1}(x) \neq 0$ and ... and $p_k(x) \neq 0$
equals $1 - r$

(r is a level of admissible fluctuations).

The statement (**PAT**$_i$) removes the semi-idealizing condition replacing it with the condition $S - i(x)_F \neq 0$ saying that at least some interferences take on values different than zero for an object x. This justifies the terminological stipulation according to which (**PAT**$_i$) is termed a probabilistic counterpart of (**AT**$_i$) and a probabilistic approximation of **T**$_i$.

The sequence of statements (**ST**$_k$), ..., (**ST**$_i$), (**SAT**$_i$), (**PAST**$_i$) is termed a semi-idealizational approximation sequence. Notice that what is usually called an econometric model might be interpreted as a semi-idealizational sequence (omitting for the sake of simplicity its approximation component). Here is the typical form of the econometric model as presented by econometricians themselves:

$$Y = f(X_1, \ldots, X_k) + r$$

where r is

a random component of the model presenting the joint effect of influence upon Y of all these factors which are not taken into account as explaining variables within the model" (Pawlowski 1964, p. 38). Its presence within the model is necessary because in the economy "causal connections occur which are not relations of pure form but their way of manifestation is disturbed by adventitious effects of inessential, accidental factors. (...) Obviously, no one is able to include all these factors into a model and only the most important variables are treated as explanatory ones. The result of joint effect of the remaining factors is taken into consideration only globally through the random component r (the symbol is changed here – L.N.). (...) r equals the difference between the value of variable Y and the function of explaining variables $X_1, \ldots, X_k$ accepted within the model. (ibid., p. 38–39)

It seems that an admissible interpretation of the quoted passages is that the above formula – equipped additionally in the condition $S - i_F \neq 0$ is a probabilistic counterpart of the semi-idealizational statement of the following, roughly, form:

$$\text{if } E - i_F = 0, \text{ then } Y = f(X_1, \ldots, X_k).$$

A more detailed analysis of the structure of it would likely reveal that it is a concretization of some initial idealizational statement referring to some of $X_1, \ldots, X_k$ variables as to the principal ones.

A probabilistic model-III will be termed the sequence:

$(\mathbf{SQ}_k, \mathbf{SC}_k)$
$(\mathbf{SQ}_{k-1}, \mathbf{SC}_{k-1})$
....
$(\mathbf{SQ}_i, \mathbf{SC}_i)$
$(\mathbf{SAQ}_i, \mathbf{SAC}_i)$
$(\mathbf{PAQ}_i, \mathbf{SAC}_i)$

determined by the relations of concretization (between statements of $\mathbf{SQ}_k, \ldots, \mathbf{SQ}_i$), consequence (between $\mathbf{SQ}_k$ and $\mathbf{SC}_k$, $\mathbf{SQ}_{k-1}$ and $\mathbf{SC}_{k-1}$, etc.), approximation (between $\mathbf{SQ}_i$ and $\mathbf{SAQ}_i$) and probabilistic approximation (between $\mathbf{SAQ}_i$ and $\mathbf{PAQ}_i$). Instead model-III in the sense employed until now will be identified as deterministic model-III. If the researcher thinks the set i_F to be empty, he is inclined to build a deterministic model. If not, he considers it to be necessary to explain phenomena by a probabilistic model.

Reconstructional and Optimizational Theories

Let us now remove condition (**D**) claiming that the explanation is the only goal of the economist. In fact, there are theories aiming only an explanation of what happens in the actual economic life, but there are also "normative models" whose goal is of a quite different nature. Let us refer to the opinions of the economists themselves:

In normative analysis, the purposes of the analysis are not limited to the empirical testing of the set of postulates, and need not even include the latter objective. The new purpose is that of recommending, to one or more of the persons or organizations represented in the analysis, a choice of course of action which can be expected to serve his or their objectives better than or at least as well as, alternative actions open to them. (Koopmans 1957, p. 134)

In some cases the term 'model' is applied to systems of equations describing situations interesting for an economist, and sometimes the term is referred to systems of equations (. . .) with requirements to find some extreme values. There are then models which limit themselves to a description of a behavior of an economic system and those which simultaneously determine an objective one might have. (Czerwinski 1979, p. 16)

I shall only outline how to include intuitions of this kind into our conceptual apparatus. Let us notice that some magnitudes are at the same time values in a given society (for a more detailed construction cf. 1980, p. 179ff). For instance, the growth of national income is an economic magnitude but usually a social value as well. One might thus distinguish statements in which the determined magnitude is at the same time a value in a given society and which say on what the optimal realization of the magnitude-value depends upon. A magnitude-value is optimally realized if realized is the extreme (minimal or maximal) case of it. Now, an optimizational idealizational statement (of the highest order) will be termed the following claim:

(**OT**$_k$) if $R(x)$ and $p_1(x) = 0$ and . . . and $p_k(x) = 0$, then $F_{extr}(x) = f_k(H(x))$

where F is a magnitude-value and $H(x)$ is an instrumental variable to be manipulated by management in order to create a value for $H(x)$ such that $F(x) = F_{extr}(x)$. An optimizational semi-idealizational statement differs from the above one only in that it possesses additionally a semi-idealizing condition. Operations of concretization, approximation, standard and probabilistic, might be defined on optimization statements quite analogically as they have been defined for standard (reconstructional) statements considered until now; the only difference is that these operations hold between optimizational statements. Optimizational models-III contain optimizational statements at least among solutions of particular systems. The fact that among economic theories there appear both reconstructional and optimizational ones[8] testifies that today theoretical economics is theoretical enough to give foundations for practical applications.

NOTES

* I am indebted to Bert Hamminga for the illuminating comments to an earlier draft of the paper. Also I would like to thank the participants of the conference "Philosophy of Economics II" (Tilburg, July 1987). Particularly, the critical remarks of Wolfgang Balzer, Theo Kuipers and Wolfgang Spohn made it possible to improve the text in many points.

1 In particular, problems of dynamics of the economic theories will not be discussed here. They are analysed in 1972 and in Nowak and Nowakowa 1978.

2 I owe the present distinction between abstraction and reduction to Zielinska's 1981 criticism of my eariler formulations of the cognitive function of idealization.

3 On the empirical methods of recognizing the significance of factors cf. Brzezinski 1977, 1978. The role of philosophical presuppositions in the idealizational theory-building is discussed in 1980, Chap. 6.

4 It is obvious that the presented grasp of the method of idealization (cf. 1970, 1971a–b,

1980) is merely one of possible approaches to this method. In some respects it resembles some alternatives (e.g., the definition of the notion of "ideal condition" and of "ideal case" in Barr 1971) in some other differs from them (cf., e.g., the role of idealizing conditions and the procedure of concretization in comparison to Suppe's 1974 approach). Cf. different approaches to idealization collected in Kuipers 1980 and a careful analysis in Niiniluoto 1985. Let us note that what Suppes 1960, 1983 terms "models of data" might be interpreted as a procedure replacing that of concretization. Instead of concretizing the theoretical statements, a purification of data is to be done so that they become capable of falling under the idealizational statements.

[5] The problematics of testing idealizational theories is omitted here. In its simplified form it is discussed in my 1980. In a more complete manner, involving a reconstruction of the procedure of operationalization of factors, it is analysed in Tuchanska 1981 and Hornowska 1982, 1985.

[6] An important simplification silently adopted here is that only properties are considered whereas relations are neglected. This results in the following: (i) predicates in the schemes of idealizational statement, its concretization etc. are monadic, (ii) they are first order-predicates, (iii) interactions between determinants of a given magnitude are neglected. Since all these effects might make, not without justification, an impression of assuming the "purely Aristotelian ontology", I would like to comment on them in short.

Re: (i). This is easy to be removed, by introducing relations and relational predicates, on the price of a complication of the conceptual apparatus from the very beginning (cf. 1977).

Re: (ii). Indeed, for some types of statements this limitation ceases to be trivial. Namely, this concerns adaptive statements (cf. 1975) which appear in different branches of science (Klawiter 1977, Lastowski 1982, Kosmicki 1985) and which involve, in an irreducible manner, predicates of higher orders (Kmita 1976, p. 41ff). The question of applicability of the operations of idealization and concretization in the realm of these statements is analyzed by Klawiter 1977, 1982 and Patryas 1979.

Re: (iii). It is presupposed here that no interactions between the elements of the space of essential factors occur. This is silently assumed already in the adopted definition of the notion of an essential property. Namely, it does not embrace the case in which two properties G and G' are non-essential for F, if taken separately, but are essential for F, if taken together. This might be expressed by saying that although G and G' are insignificant for F, their interaction int(G, G') is essential for F. Therefore, the essential structure for F is, in Brzezinski's 1975 terms, of the second kind (if there is, additionally, a single factor H essential for F) or of the third kind (if the interaction is the only element influencing F). I have not included these problematics because it would complicate our schemes very much. Moreover, the methodological significance of accounting for interactions is controversial (cf. Brzezinski 1977, my 1977, Gaul 1985).

[7] This understanding of the term "approximation" corresponds to the meaning (b) of the term in the list analysed in Moulines 1976, p. 203ff. Let us add that Kuhn's "paradigmatic examples" (e.g., Kuhn 1976, p. 182), so important for the structuralist concept of theory (e.g., Sneed 1976, p. 120, Niiniluoto 1985, p. 144) can be interpreted within the proposed conceptual apparatus as those cases that confirm already the approximation of the first chain in a given idealizational theory:

$(\mathbf{Q}_k, \mathbf{C}_k)$,
$(\mathbf{Q}_{k-1}, \mathbf{C}_{k-1})$
....
$(\mathbf{Q}_i, \mathbf{C}_i)$
$(\mathbf{A}_i, \mathbf{AC}_i)$,

i.e., those which satisfy statements of $(\mathbf{AQ}_k, \mathbf{AC}_k)$. For the set of cases satisfying statements of $(\mathbf{AQ}_j, \mathbf{AC}_j)$ is properly included in the set satisfying statements of $(\mathbf{AQ}_{j+1}, \mathbf{AC}_{j+1})$. "Paradigmatic examples" are then those which belong to all the sets at the same time.

[8] Contrary to a supposition of Haendler 1980 who claims that the only criterion of "normative models" might be whether the scientific community is inclined to employ the given conceptual structure "at formulating concrete economic policies" (p. 154), it seems that there is a (quasi-) structural criterion allowing to distinguish optimizational models. "Quasi-", since the notion of value calls for a relativization to sociological notions. Let us also comment on another interesting claim put forward by this author. He distinguishes between an "empirical theory (which) is characterized by the claim that the entities forming its models are part of the ontological inventory of the actually existing world" and a "pure theory (that) does not intend to speak about reality. A pure theory is just a (...) picture of a possible world which does not actually exist" (Haendler 1982, pp. 74–75). Yet, even the structure of scientific magnitudes seems to exclude any "empirical theory" in this meaning of the word. For every continuous magnitude contains some classes of abstraction (Ajdukiewicz's 1974 construction of a magnitude, generalized by Wojcicki 1979, as a family of the equivalence classes is presupposed here) that are likely empty in the actual world, e.g., 36,6677.... Celsius degrees, and the only inhabitants of them might be possibilia only. On the other hand, if model-III give a proper view on the basic structure of economic theories one might say that every one of them "intends to speak about reality". Statements of $(\mathbf{Q}_k, \mathbf{C}_k), \ldots, (\mathbf{Q}_i, \mathbf{C}_i)$ in fact do not speak about the actual world for they refer to (less and less) idealized worlds. Yet, the next chain in model-III, i.e. $(\mathbf{AQ}_i, \mathbf{AC}_i)$, is to speak about the real world (among other possible worlds). If this grasp is tenable, there would be no distinction between the "empirical" and "pure" theories but in every (idealizational) theory there would be a "pure" and an "empirical" part.

REFERENCES

Ajdukiewicz, K.: 1974, *Pragmatic Logic*, Reidel, Dordrecht, Boston.

Balzer, W. and M. Heidelberger (Hrsg.): 1983, *Zur Logik empirischer Theorien*, de Gruyter, Berlin, New York.

Barr, W. F.: 1971, 'A Syntactic and Semantic Analysis of Idealizations in Science', *Philosophy of Science* **38**, 258–72.

Brzezinski, J.: 1975, 'Interaction, Essential Structure, Experiment', *Poznan Studies in the Philosophy of the Sciences and the Humanities*, vol. 1, no. 1, Gruener, Amsterdam, pp. 43–58.

Brzezinski, J.: 1977, *Metodologiczne i psychologiczne wyznaczniki postepowania badaw-*

czego w psychologii (Methodological and Psychological Determinants of the Research Process in Psychology), Poznań Univ. Press, Poznań.

Brzezinski, J.: 1978, 'Odkrycie i uzasadnienie (na przykladzie budowy ekonometrycznego modelu wielokrotnej regresji liniowej)' (Discovery and Justification (on the example of the econometric model of the multiple linear regression)), in P. Buczkowski and L. Nowak 1978, pp. 11–30.

Brzezinski, J. (ed.): 1985, *Consciousness: Methodological and Psychological Approaches*, *Poznan Studies in the Philosophy of the Sciences and the Humanities*, vol. 8, Rodopi, Amsterdam.

Buczkowski, P. and A. Klawiter (eds.): 1985, *Klasy, swiatopoglad, idealizacja* (Classes, the Worldlook, Idealization), *Poznanskie Studia z Filozofii Nauki* **9**, PWN, Warszawa, Poznań.

Buczkowski, P. and L. Nowak (eds.): 1978, *Teoria ekonomiczna: metodologia i rekonstrukcje* (The Economic Theory: Methodology and Reconstructions), Poznań Academy of Economics Press, Poznań.

Czerwinski, Z.: 1979, *Matematyka na usługach ekonomii* (Mathematics Serving the Science of Economy), 4th ed., PWN, Warszawa.

Domar, E. D.: 1962, *Essays in the Theory of Economic Growth* (after the Polish translation, PWN, Warszawa).

Friedman, M.: 1953, *Essays in the Positive Economics*, University of Chicago Press, Chicago.

Gaul, M.: 1985, 'The Problem of Interaction in the Protoidealizational Model of the Investigative Process. Modifications and Developments', in J. Brzezinski (ed.), pp. 58–80.

Haendler, E. W.: 1980, 'The Role of Utility and of Statistical Concepts in Empirical Economic Theories: the Empirical Claims of the Systems of Aggregate Market Supply and Demand Functions Approach', *Erkenntnis* **15**, 129–157.

Haendler, E. W.: 1982, 'The Evolution of Economic Theories. A Formal Approach', *Erkenntnis* **18**, 65–96.

Hintikka, J. and F. Vandamme (eds.): 1985, *Logic of Discovery and Logic of Discourse*, *Communication & Cognition*, vol. 18, no. 1/2.

Hornowska, E.: 1982, 'O problemach operacjonalizacji w idealizacyjnej teorii nauki' (On Problems of Operationalization in the Idealizational Concept of Science), *Studia pedagogiczne* **XLIV**, 65–86.

Hornowska, E.: 1985, *Operacjonalizacja wielkosci psychologicznych w ramach idealizacyjnej koncepcji nauki* (Operationalization of Psychological Magnitudes in the Framework of the Idealizational Concept of Science), unpublished Ph.D. thesis, Poznań.

Klawiter, A.: 1978, *Problem metodologiczenego statusu materializmu historycznego* (The Problem. of the Methodological Status of Historical Materialism), PWN, Warsaw, Poznań.

Klawiter, A.: 1982, 'Adaptacja i konkurencja. Przyczynek do klasyfikacji zaleznosci adaptacyjnych' (Adaptation and Competition. A Contribution to the Classification of the Adaptive Dependencies), in K. Lastowski and J. Strzalko (eds.), pp. 81–100.

Kmita, J.: 1976, *Szkice z teorii poznania naukowego* (Essays in the Theory of Scientific Cognition), PWN, Warsaw.

Kmita, J. (ed.): 1977, *Zalozenia teoretyczne badan nad rozwojem historycznym* (Theoretical Assumptions of the Investigation of the Historical Development), PWN, Warszawa.
Kosmicki, E.: 1985, *Biologiczne koncepcje zachownia* (Biological Theories of Behaviour), PWN, Warsaw, Poznań.
Koopmans, T. C.: 1957, *Three Essays on the State of Economic Science*, McGraw-Hill, New York, Toronto, London.
Kuipers, T. A. F. (ed): 1980, *Idealizatie en konkretisering*, Groningen: Filosofisch Instituut, Rijksuniversiteit.
Kuhn, T. S.: 1976, 'Theory-Change as Structure-Change: Comments on the Sneed Formalism', *Erkenntnis* **10**, 179–200.
Loux, M. J.: 1983, 'Recent Work in Ontology', in K. G. Lucey and T. R. Machan (eds.), pp. 3–38.
Lastowski, K.: 1982, 'The Theory of Development of Species and the Theory of Motion of Socio-economic Formation', in L. Nowak (ed.), pp. 122–157.
Lastowski, K. and J. Strzalko (eds.): 1982, *Filozofia i biologia* (Philosophy and Biology), PWN, Warsaw, Poznań.
Lucey, K. G. and T. R. Machan (eds.): 1983, *Recent Work in Philosophy*, Rowman and Allanheld, Totowa, N. J.
Moulines, U. C.: 1976, 'Approximate Application of Empirical Theories: A General Explication', *Erkenntnis* **10**, 201–227.
Niiniluoto, I.: 1983, 'Theories, Idealizations, Approximations', paper read at Salzburg 7th Congress of Logic, Methodology and Philosophy of Science (quoted after the Polish translation in: Buczkowski/Klawiter 1985, pp. 115–162).
Niiniluoto, I.: 1985, 'Paradigms and Problem-Solving in Operations Research', in J. Hintikka and F. Vandamme (eds.), pp. 141–155.
Nowak, L.: 1970, 'O zasdzie abstrakcji i stopniowej konkretyzacji (On the Principle of Abstraction and Gradual Concretization)', in collective work, *Zalozenia methodologiczne 'Kapitalu' Marksa* (The Methodological Assumptions of Marx's 'Capital'), KiW, Warszawa, pp. 123–217.
Nowak, L.: 1971a, *U podstaw Marksowskiej metodologii nauk* (Foundations of Marxian Methodology of Science), PWN, Warszawa.
Nowak, L.: 1971b, 'The Problem of Explanation in Carl Marx's 'Capital'', *Quality and Quantity*, vol. V, no. 1, pp. 311–338.
Nowak, L.: 1972, *Model ekonomiczny. Stadium z metodologii ekonomii politycznej* (Economic Models. A Study in the Methodology of the Political Economy), PWE, Warszawa.
Nowak, L.: 1977, 'Odpowiedz (A Reply)', in Kmita, (ed.), pp. 71–83.
Nowak, L.: 1980, *The Structure of Idealization. Towards a Systematic Reconstruction of the Marxian Idea of Science*, Reidel, Dordrecht, Boston, London.
Nowak, L. and I. Nowakowa: 1978, 'Pewne osobliwosci rozwoju teorii ekonomicznych' (Some Peculiarities of the Development of Economic Theories), in P. Buczkowski and L. Nowak (eds.), pp. 62–73.
Patryas, W.: 1979, *Idealizacyjny charakter interpretacji humanistycznej* (The Idealizational Nature of the Humanitic Interpretation), Poznań Univ. Press, Poznań.
Pawlowski, Z.: 1964, *Wstep do ekonometrii* (Introduction to Econometrics), Warszawa, PWE.

Popper, K. R.: 1959, *The Logic of Scientific Discovery*, Hutchinson, London.
Suppe, F.: 1974, 'Theories and Phenomena', in W. Leinfellner and E. Koehler (eds.), pp. 45–91.
Suppes, P.: 1960, 'A Comparison of the Meaning and Uses of Models in Mathematics and the Empirical Sciences', *Synthese* **12**, 287–301.
Suppes, P.: 1983, 'Modelle von Daten', in W. Balzer and M. Heidelberger (eds.), pp. 191–204.
Sneed, J. D.: 1976, 'Philosophical Problems in the Empirical Science of Science: A Formal Approach', *Erkenntnis* **10**, 115–146.
Tuchanska, B.: 1981, *Czynnik – wielkosc, zwiazek – zaleznosc* (Factor – magnitude, regularity – dependence), PWN, Warszawa, Poznań.
Wojcicki, R.: 1979, *Topics in the Formal Methodology of Empirical Sciences*, Reidel, Dordrecht, Boston, London.
Zielinska, R.: 1981, *Abstrakcja, idealizacja, generalizacja* (Abstraction, Idealization, Generalization), Poznań Univ. Press, Poznań.

Manuscript received 25 January, 1988

University of Groningen
Department of Philosophy
Westersingel 19
NL-9700 AV Groningen
The Netherlands

BERT HAMMINGA*

SNEED VERSUS NOWAK: AN ILLUSTRATION IN ECONOMICS

This article deals with Nowak's reconstruction of Marx's price theory in volume III of *Capital*. It is claimed that Marx's analysis indeed leads to a series of idealizational statements in Nowak's sense, to which an analysis is added of the logical procedure by which Marx successively arrives at each of the idealizational statements. This procedure explains the structure found by Nowak. Nowak's analysis should thus be supplemented by a "structuralist" consciousness of Marx's theory core, and structuralists can learn from Nowak's results that the claims of Marx's theory are empirically vacuous, and hence cannot form Marx's prime purpose for constructing the theory.

It is the purpose of this article to discuss Nowak's method of describing the skeleton of economic theories and compare it with the Sneed Stegmüller Balzer type of reconstruction, known as the "structuralist" approach. I claim we learn about economics by doing so. For Nowak's approach, I refer to his book *The Structure of Idealization* (1980). His best known reconstruction is that of the law of value in volume III of Marx's "Capital", but he also devoted attention to volume I of "Capital", to some modern theories of economic growth, and to econometrics.

In the framework of the structuralist approach, we have several reconstructions of neoclassical theories (by Händler (1980) Balzer (1982), (1985) Haslinger (1982), (1983) and Hands (1985), that were integrated by Kuipers and Janssen in this volume, and one by Hamminga and Balzer (1986) that includes production). But we also have structuralistic reconstructions of parts of Marx's "Capital", by Diederich and Fulda (1978), Diederich (1982) and García de la Sienra (1982).

Finally, Sneed (1976) has already made thoughts about a reconstruction of his own method of reconstruction, and so did Nowak (1980, p. 111 etc. and in this volume).

So, what I could do, is make a Sneedian reconstruction of Nowak. Or a Nowakian reconstruction of Sneed. Or, I could even try to reconstruct them both, using an even more general theory of reconstruction (Niiniluoto, 1983).

Erkenntnis **30** (1989) 247–265.

But, since we learn much better from examples than from the abstract comparison of metatheories, and since Marx's "Capital" is more or less common ground, I will try to make my observations in connection with Marx's treatment of the law of value, in volume III of "Capital", which I consider to be a very fine and not really eccentric example of the average theory structure you find in economics, though it has some peculiarities. I will not mix up the *reconstruction* with an *evaluation* of that theory. It is irrelevant whether or not the reconstructor believes the theory to be true, and it is not even relevant for this article whether or not Marx believed this himself.

Marx's fundamental law of value says that for any two types of commodities x and y:

$$\frac{p(x)}{p(y)} = \frac{w(x)}{w(y)}$$

i.e., price ratio of the commodities x and y, $p(x)/p(y)$, equals the ratio of the labour times required in their respective production $w(x)/w(y)$ (also called the ratio of their *values*).

Choosing $p(y)$ and $w(y)$ as units of account, the law reads: for all commodities x:

$$p(x) = w(x)$$

This version of the law says that the amount of x that exchanges for one unit of y equals the time necessary for the production of one unit of x (measured in time units equal to the time necessary for the production of one unit of y). This means that $p(x)$ is determined by the composition of $w(x)$. For the understanding of this composition Marx appeals to the reader's power of imagination: he hopes that all imaginable production processes (only some of which correspond to really existing production processes) can be represented in such a way that $w(x)$ consists of three parts, as in the following illustration:

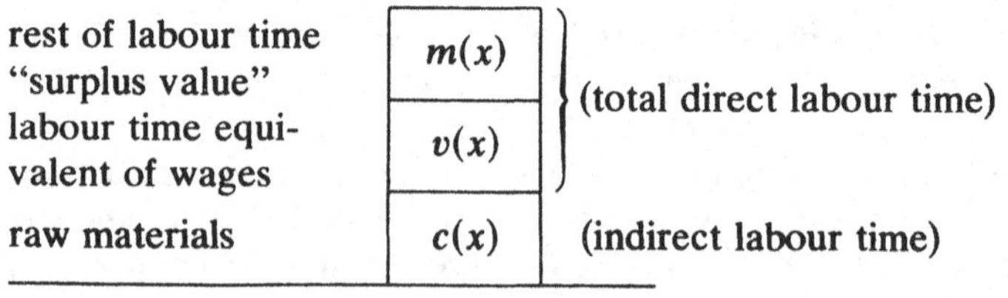

Thus, $w(x)$ is defined by $w(x) := c(x) + v(x) + m(x)$, reflecting the following intuitions:

(1) It takes time to produce x.
(2) You need raw materials.
(3) It usually has taken time to produce raw materials.
(4) You have to pay the workers.

And so, the value $w(x)$ of x consists of the indirect labour time $c(x)$ of the raw materials that go into x, and the direct labour time $v(x) + m(x)$ that workers need to produce x. Only part of this direct labour time is needed to pay the workers' wages. That part is $v(x)$ (so called *variable capital*). What is left is $m(x)$, the *surplus value* that you can at this stage identify with profit. So, we know $w(x)$: it is the sum of these three amounts of labour time.

As Nowak pointed out, and as is very clear in *Capital*, Marx never states that there actually exist or have existed economies satisfying:

$$\forall_{x \in E} \, (p(x) = w(x))$$

where E is the set of all types of goods produced in such an economy. For Marx, the expression above is an *ideal law* that only holds for abstract, non existing economies satisfying the following conditions:[1]

(C_1) All commodities of the same type require equal labour time.
(C_2) All types of commodities need c (raw materials) and v (labour time equivalent of wages) in the same proportion ($c(x)/v(x)$ is the same for all x).
(C_3) There is no fixed capital.
(C_4) The turnover time of every capital is one ("period of analysis").
(C_5) There is no merchant capital.
(C_6) The purely financial sphere (banking) receives no surplus value (there is no interest payed).
(C_7) There is no differential ground rent.
(C_8) There is no absolute ground rent.

Nowak's reconstruction T_k of Marx's fundamental idealizational statement is written:

(T_k) If E is the set of goods produced in some real capitalist economy and (C_1) and (C_2) and ... and (C_8) then $\forall_{x \in E}\,(p(x) = w(x))$

For purposes of exposition the E-clause shall be suppressed and we adopt the following notation and terminology:

$$(T_k) \qquad \bigwedge_{1}^{8} C \rightarrow \forall_{x \in E}\ (p(x) = w(x))$$

where $\bigwedge_{1}^{8} C$ is a shorthand for (C_1) and (C_2) and ... and (C_8).

These 8 conditions are called *idealizing conditions*. The total expression T_k is called an *idealizational statement*. It is a material implication. The consequent of this implication is called an *ideal law*.

With respect to the set of ideal conditions, we can observe the following:

(1) They are all at variance with our image of the capitalist economy (it is a common sense belief that they are false and Marx is, of course, very well aware of this).

(2) There is a definite reason for the introduction of every one of them.

(3) Nowak does not *formally* reconstruct these reasons, though in analysing Marx's succession of modifications of the law "of value", Nowak uses these reasons in order to explain the exact form of the modified laws. That is, these reasons can not be found by inspecting the idealizational statement. They come from "somewhere else".

So, now I am going to give two fully specified examples of these reasons, for the first two of the conditions, and an outline of what these reasons are for the other conditions.

(C_1) is *necessary* under competitive conditions on the commodity markets.

Proof. If commodities of the same type would not require equal labour time, then they would have to sell at different prices (if the fundamental law of value is to apply). But if there is a free market for these commodities, this market would equalize prices. Therefore, the fundamental law of value can only apply under competitive conditions if commodities of the same type require equal labour time. Q.E.D.

The important thing to note is that the necessity of ideal condition

(C_1) is ***proven*** from competitive conditions. And this is the same for (C_2) which is the condition that gave rise to the notorious transformation problem.

The condition (C_2) that all types of commodities need c and v in the same proportion is necessary under competitive conditions on the capital market and labour market.

Proof. (1) If capital freely competes, then the rate of profit $p'(x)$ in the industry x should be the same for all x (differences in such profit rates would cause capital to move and to be reinvested in the most profitable industries). If $p(x) = w(x)$, the rate of profit on x can be expressed as

$$p'(x) = \frac{m(x)}{c(x) + v(x)}$$

(2) The rate of exploitation $m'(x) = m(x)/v(x)$ must be the same for all industries x. If not, the worker's wages would differ over industries, which would cause them to move to the industry that pays the highest wages, until wage would again be equal everywhere. (3) If, however both $p'(x)$ and $m'(x)$ should not differ over industries, then the "organic composition of capital" $c(x)/c(x) + v(x)$ cannot differ over industries, because by definition

$$\frac{c(x)}{c(x) + v(x)} = 1 - \frac{m'(x)}{p'(x)}$$

Therefore $c(x)/c(x) + v(x)$ must be the same for all x, hence also $c(x)/v(x)$. Q.E.D.

(C_3) Introduction of fixed capital *logically* requires strong conditions concerning the distribution of fixed capital over the industries if profits are to remain equal everywhere, and if you want the products to be sold at their value. Relaxation of these strong conditions under maintenance of profit equality everywhere, *logically* leads to price deviations from values, indeed, completely according to Nowak's idea of concretisation.

(C_4) Turnover time differences can be ***proven*** to have the same effects as disproportionalities in fixed capital, if profit rates are to remain equal everywhere.

(C_5) Merchant capital is improductive, according to Marx, and if merchants take surplus value from the industry, and if they do not do so in very nicely suiting proportions they can be *proven* to cause additional price deviations if profit rates are to remain equal everywhere.

(C_6) The same holds for bankers.

(C_7) and (C_8) If there is a supply monopoly of land, the free competition between land hiring capitalist-farmers and competitive conditions on the markets for commodities produced with the help of land can be *proven* jointly to cause different types of land-rent, which in turn can be *proven* to cause different types of price deviations.

Marx's strategy in "Capital", as Nowak pointed out, consists of removing successively these idealizing conditions, at the same time introducing a modification of the ideal law such that the modified form is not inconsistent with competitive conditions in situations where the ideal condition does not hold.

The analysis of ideal conditions above draws attention to the fact that behind this Nowak-series of idealizational statements

$$\mathrm{T}_k, \mathrm{T}_{k-1}, T_{k-2}, \ldots .$$

there is a coherent and uniform theory-core K that contains the main assumptions from which each of the ideal conditions occuring in T_i is *proven logically*. The axioms in this theory core are that competitive conditions rule on the *capital market*, the *labour market* and on the *markets for every type of commodity*.

So, I disagree with Nowaks attempt, in this volume, to assign to *deduction* solely the purpose of drawing consequences from idealizational statements. I think deduction is the essence of economics and that the ideal conditions occurring in the idealizational statements, the structure and sequence of which is so marvelously reconstructed by Nowak, are in fact *themselves* deduced. Deduced from the law of value in conjunction with theory core K that is, from the assumption of labour time determining prices and of competitive conditions ruling everywhere. You can *prove logically* that this conjunction is only consistent if the ideal conditions hold.

The dropping of Condition 1, Condition 2, etc. forces Marx *logically* to modify something somewhere in this conjunction and he chooses to do so by modifying the determination of prices and profits, under maintenance of competitive conditions, everywhere.

Logically, Marx could have stuck to his labour time determined prices! Then he would have been forced logically to accept profit rate differences, or differences in the rate of exploitation, or differences in prices of commodities of equal type. Marx considered it to be his job to keep the conjunction (of the law of value with the assumption of competitive conditions ruling everywhere) consistent under dropping of his special conditions. In doing so, he *keeps* the competitive assumptions, from the deep to the surface. But if you look for prices purely determined by labour time, you find them only in the deep. That is: under all ideal conditions. Going upwards, additional forces come in and cause "deviations". So, I think, Marx considered competitive conditions to be more essential than labour time, for the formation of profits and prices. And he says so! In the pre-capitalist economics, he writes, the whole price determination by labour time was impossible because of the absence of competitive conditions.[2] Competitive conditions form Marx's basic theory of the capitalist economy. They make possible a system of exchange ratio's between commodities purely determined by labour time, though only under conditions that are never satisfied in practice. This is the logic of the problem situation in which Marx finds himself:

$$(L_1) \qquad K \textit{ and } \forall_{x \in E} (p(x) = w(x)) \Rightarrow \bigwedge_1^8 C$$

Read: from competitive conditions obtaining on all markets (K) and commodities being sold at their values it *follows logically* that (C_1) and (C_2) and . . . and (C_8). This expression is equivalent to

$$(L_2) \qquad K \Rightarrow \forall_{x \in E} (p(x) = w(x)) \rightarrow \bigwedge_1^8 C$$

Read: under competitive conditions it can be *proven logically* that, if commodities sell at their values, then conditions (C_1), (C_2), . . . , (C_8) must be satisfied.

Now, let us approach Nowak's arrangement of expressions more closely by introducing yet another equivalent report of this Marxian proof:

$$(L_3) \qquad K \Rightarrow \neg \bigwedge_1^8 C \rightarrow \neg \forall_{x \in E} (p(x) = w(x))$$

Read: under competitive conditions it can be proven logically that if at least one of the ideal conditions is *not* satisfied, then *not* all commodities will sell at their values. Let us call the proven material implication a *competitive corollary*.

Let us now compare this to Nowak's T_k (saying that if the ideal conditions hold, then the commodities will sell at their values):

$$(\mathrm{T}_k) \qquad \bigwedge_{1}^{8} \mathrm{C} \rightarrow \forall_{x \in E} (p(x) = w(x))$$

As Nowak clearly pointed out, such an idealizational statement is not logically empty, but it has no testable content, since the testing conditions for the law ((C_1) to (C_8)) are never satisfied in real capitalist societies, as everybody knows.

The following truth table clarifies the difference:

	$\bigwedge_1^8 \mathrm{C}$	$\forall x(w(x) = p(x))$	competitive corollary $\neg(\bigwedge_1^8 \mathrm{C}) \rightarrow \neg\forall x(p(x) = w(x))$	idealizational statement $\bigwedge_1^8 \mathrm{C} \rightarrow \forall x(p(x) = w(x))$
I	1	1	11	
II	1	0	1	0
III	0	1	0	1
IV	0	0	1	1

Nowak, Marx, and probably the reader, too, all agree: all capitalist countries are examples of case IV of this truth table: the ideal conditions do not hold, and prices do not sell at their values. Nowak's interpretation of Marx's statements pertaining to the law of value is that Marx expresses the following statement: if the ideal conditions would actually hold somewhere, that is, if we would have a society that is an example of case I or II, then this society would be an example of case I (and not an example of case II). But the least we can say of Marx is that he actually proves, proves logically, that competitive conditions rule out case III; they imply the *logical* impossibility of a society where commodities sell at their values while at the same time the ideal conditions are not all satisfied. Such a case would logically falsify the competitive conditions, that is, Marx would conclude from the occurrence of such a case that competitive conditions must have been violated. But not only Marx, because this is a purely logical affair. It is the affair of deduction. So, the least we can say is, that Marx *deduced* his ideal conditions, from competitive conditions.

And, one could say, he is not at all happy about having deduced these ideal conditions, because he had to get rid of them in order to construct a theory of value determining price under competitive conditions, that does not so obviously rule out all real capitalist economies. And this is what a large part of volume III of "Capital" is all about. His course of analysis can be reconstructed as consisting of the following 27 steps (to be explained below).

(1) Initial implication: $K \rightarrow \underset{x \in E}{\forall} (p(x) = w(x))$

(2)(L_0) Competitive corollary: $K \Rightarrow \neg \bigwedge_{1}^{8} C \rightarrow \neg \underset{x \in E}{\forall} (p(x) = w(x))$

(3)(T_0) Idealizational statement: $(K \text{ and } \bigwedge_{1}^{8} C) \rightarrow \underset{x \in E}{\forall} (p(x) = w(x))$

(4) Construction of f_1 (necessary conditions: f_1 should make the claims independent of C_1 and $\sum_x q_x \cdot f_1(w(x), \ldots) = \sum_x q_x \cdot w(x)$)

(5)(L_1) Competitive corollary:

$$K \Rightarrow \neg \bigwedge_{2}^{8} C \rightarrow \neg \underset{x \in E}{\forall} (p(x) = f_1(w(x), \ldots))$$

(6)(T_1) Idealizational statement:

$$(K \text{ and } \bigwedge_{2}^{8} C) \rightarrow \underset{x \in E}{\forall} (p(x) = f_1(w(x), \ldots))$$

(7) Construction of f_2 (necessary conditions: f_2 should make the claims independent of (C_2) and $\sum_x q_x \cdot f_2(w(x), \ldots) = \sum_x q_x \cdot w(x)$)

(8) Competitive corollary:

$$K \Rightarrow \neg \bigwedge_{3}^{8} C \rightarrow \neg \underset{x \in E}{\forall} (p(x) = f_2(w(x), \ldots))$$

.

.

.

(23)(L_7) Competitive corollary:

$$K \Rightarrow \neg C_8 \rightarrow \neg \underset{x \in E}{\forall} (p(x) = f_7(w(x), \ldots))$$

(24)(T_7) Idealized statement:

$$(K \text{ and } C_8) \rightarrow \underset{x \in E}{\forall} (p(x) = f_7(w(x), \ldots))$$

(25) Construction of f_8 (necessary conditions: f_8 should make the claims independent of (C_8) and $\sum_x q_x \cdot f_8(w(x), \ldots) = \sum_x q_x \cdot w(x)$)

(26)(L_8) Competitive corrollary: $\forall_{x \in E} (p(x) = f_8(w(x), \ldots)$ is consistent with K

(27)(T_8) Idealizational statement: $K \rightarrow \forall_{x \in E} (p(x) = f_8(w(x), \ldots))$

The whole exposition of *Capital* somehow starts with the idea (line 1 in the scheme above) that under the competitive conditions of a capitalist society, commodities sell at their values, but, right from the beginning, Marx concedes that this is not happening in reality. The rhetorical equipment used by Marx is the concept of value. In *Capital*, Marx uses the notion "law of value" ("Wertgesetz") in an ambiguous way. After having given only four pages of considerations on the first four pages of Capital, Marx concludes (Marx 1867, p. 54): "therefore the amount of socially necessary labour . . . determines the amount of value". This is called later on (Marx 1867, p. 202) "law of value": "according to the general law of value are, for instance, 10 pounds of yarn equivalent to 10 pounds of cotton . . . if the same labour time is necessary . . .". This is, to my knowledge, the first time Marx uses the term "Wertgesetz". There is absolutely no mention of price, it concerns the relation between "value" and "socially necessary" labour time. This is repeated on Marx (1867), p. 337 where Marx speaks about "the law of the determination of value by labour time".

Is equating value to socially necessary labour time meant as a *law* or as a *definition* of the concept of value? In my opinion Marx is unclear about this. Moreover I think this unclearness is deliberate, firstly because in handling many concepts in Capital, Marx reveals perfectly to understand the difference between definitions and statements. Secondly: if Marx would consider it to be a definition, then why call it a law? But if he considers it as a law, then he should have provided separately a definition of value. Thirdly: Marx knows very well that the reader is tempted to identify "value" with "price" or "natural price" if confronted with statements like "the amount of socially necessary labour determines the amount of value". Marx (1867, p. 557) mentions that the workers' money wage is often referred to as the value of labour ("Man spricht hier vom Wert der

Arbeit"). (This, however, says Marx, is false! It is false because we now "know" that value is something different, that value is incorporated labour and that it does not make sense to say that labour *has* value. It *is* value).

Whatever Marx says about value, he carefully avoids, right from the beginning, to equate it unconditionally with price. "Amounts of value" have a "money expression" (Marx 1867, p. 116) ("Geldausdruck") which, already here, in volume I, is "more or less", dependent upon "given circumstances" (Marx 1867, p. 117) that are not yet specified.

Yet, and obviously, the reader of "Capital" is (I presume: deliberately) put into the temptation to regard values of commodities as "first approximations" of their price, and, through all modifications in Volume III, the sum total of the prices of all commodities remains equal to the sum total of all values produced, and therefore the average price of all commodities remains equal to the average value of commodities. Nevertheless, *individual* values are not all equal to *individual* prices, unless (C_1) to (C_8) are satisfied, and this is what, in the second step, (L_0) says. (L_0) describes a mathematical state of affairs and therefore does not involve any hypothesis, be it on real or on abstract economies. The idealizational statement[3] T_0, on the contrary, is an hypothesis, though it is one about an abstract economy, and hence cannot be tested. T_0 cannot be deduced from L_0 and the same holds for all transitions from (L_i) to (T_i). Marx deduces his ideal conditions but he does not at all deduce his idealizational statements (T_i). This consistently performed "Marxian jump" from L_i to T_i has something puzzling about it, though it is a jump that many will, I think, be familiar with: if you find that one of your favourite conclusions does not hold if some conditions are violated you are inclined to maintain your conclusion, gnashing your teeth while conceding that the said conditions must be satisfied. In the truth table shown above you can see that the competitive corrollary allows case II as a logical possibility: in some capitalist economy the ideal conditions are satisfied but commodities do not sell at their values. Part 1 of the "Marxian jump" to the idealizational statement consists of excluding this case (the idealizational statement has a zero truth value for case II). Part 2 of the jump, if we stick to Nowak's reconstruction, consists of removing the zero for case III. But case III is a logical impossibility, given the competitive conditions that form Marx's very reason to introduce the ideal conditions. There may be reason here to consider replacing the material implication in Nowak's idealizational statement by material equivalence, which would remove part of the "Marxian jump", thereby retaining the zero that Marx logically discovered for case III (the material equivalence would yield a column [1001]). I will not consider this here and leave the "Marxian jump" as it is; a zero for case III "moves up" to case II of the truth table.

Let us get back to step (3) of the scheme above. After the jump, Nowak constructs a function f_1 of value and other factors. This function should replace the simple expression $w(x)$ in the law of value. The first, and main requirement for the new function is that it should not anymore involve (C_1) as a necessary condition. Secondly, f_1 should be such that if (C_1) where satisfied, the value of f_1 should be equal to $w(x)$, that is, in that case the abstract law of value as expressed on the first line of the scheme above, should hold. These two requirements make the claims independent of (C_1). Thirdly, f_1 should be economically plausible. What is, and what is not plausible cannot be fully specified logically, because it depends upon the specific economic force introduced by dropping an ideal condition. In case of (C_1), for instance, this involves the introduction of an average labour time required to produce x, which determines the value of x. Capitals producing x somewhat faster will sell x above individual value and yield surplus profit, capitals producing x slower will sell x below individual value and yield less than the average rate of profit. There is, however, one requirement of plausibility that Marx retains over all modifications: the sum of all prices $q_x \cdot f_i(w(x), \ldots)$ (where q_x is the amount of commodity x that is produced) should remain equal to the sum of all values $q_x \cdot w(x)$, and hence the sum of all profits should remain equal to the sum of all surplus value.[4] That is: every modification f_i in the law of value involves a new way of *redistributing* the original value of commodities to arrive at prices different from these original values.[5]

It can be seen in the scheme above that the procedure leading to f_1 in step 4 is now iterated seven times, finally yielding the idealizational statement (T_8) that does not anymore contain any of the ideal conditions (C_1) to (C_8). The relationship between any f_i and f_{i-1} in *Capital* can be described generally as follows:

(1) There is an ideal condition C_i that is implied by *K and*

$\forall_{x \in E} (p(x) = f_{i-1}(w(x), \ldots)$, and that is *not* implied by *K and* $\forall_{x \in E} (p(x) = f_i(w(x), \ldots))$.

(2) *If* C_i *then* $f_{i-1}(w(x), \ldots) = f_i(w(x), \ldots)$.

(3) $\sum_{x \in E} q_x \cdot f_{i-1}(w(x), \ldots) = \sum_{x \in E} q_x \cdot f_i(w(x), \ldots)$, where q_x is the quantity of x that is produced.

(4) Apart from (3) f_i and f_{i-1} are economically plausible on ad hoc criteria.

Requirement (1) says, in Nowak's terms, that the idealizational statement containing f_i should be a *concretization* of the idealizational statement containing f_{i-1}. Requirement (2) could properly be called a *reducibility requirement*. These two requirements guarantee logical independency of (C_i). Requirement (3) makes sure the law of value remains true for the aggregate sum of value and price (and so for the average). In addition to these three requirements, every special modification has its own specific criteria of plausibility, for which, of course, a general logical description is not possible. This is reflected in requirement (4).

What would Nowak's achievement be if described in the language of the Sneed–Stegmüller–Balzer-type of structuralism? I wish to maintain that Nowak did not reconstruct Marx's *theory* in the structuralist sense of the word. He did, quite successfully, reconstruct the sequence of claims that Marx obtains from his theory by deduction and jump. Idealizational statements should be looked upon as the claims of a theory.

And so, structuralism could learn from Nowak:

(1) The claims of Marx's theory, Nowaks idealizational statements T_i, have no testable content, because they have counterfactual conditions.

(2) They are made so on purpose.

(3) There is not one claim, but a whole series of them.

(4) There are very clear logical relationships between the claims in such a series. Roughly, each next claim has less counterfactual conditions then the previous one.

Let me illustrate the situation with the Venn diagrams below in Figures 1 and 2: In Figure 1, I depict P_0 (M_{pp}), the power set of the set

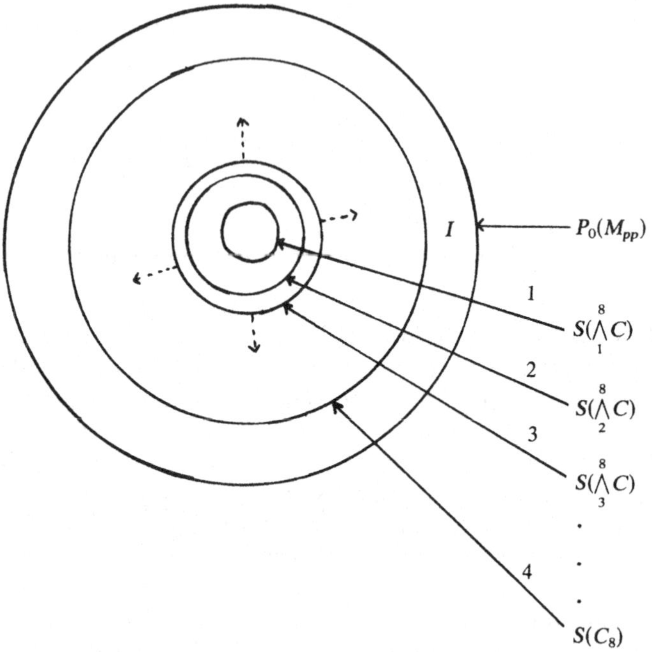

Fig. 1. $P_0(M_{pp})$ The set of all sets of imaginary capitalist economies.

of all imaginary capitalist economies. For pointing at the crucial features, I do not even have to specify the sets and functions that are involved.

By S(C) I denote the set for the elements of which the condition C holds.

By way of exercise I analyze, in Figure 1, the ideal conditions for their own sake. The smallest set, indicated by arrow 1, is the set of sets of imaginary capitalist economies for which all 8 ideal conditions hold. The set indicated by arrow 2 is the set of sets for which all ideal conditions hold except for (C_1). The set indicated by arrow 3 is the set of sets for which all ideal conditions hold except for (C_1) and (C_2), etc., until the set indicated by arrow 4. This is the set of sets of imaginary capitalist economics that satisfy (C_8) only.

Now, the set I of imaginary capitalist economies, the structure of

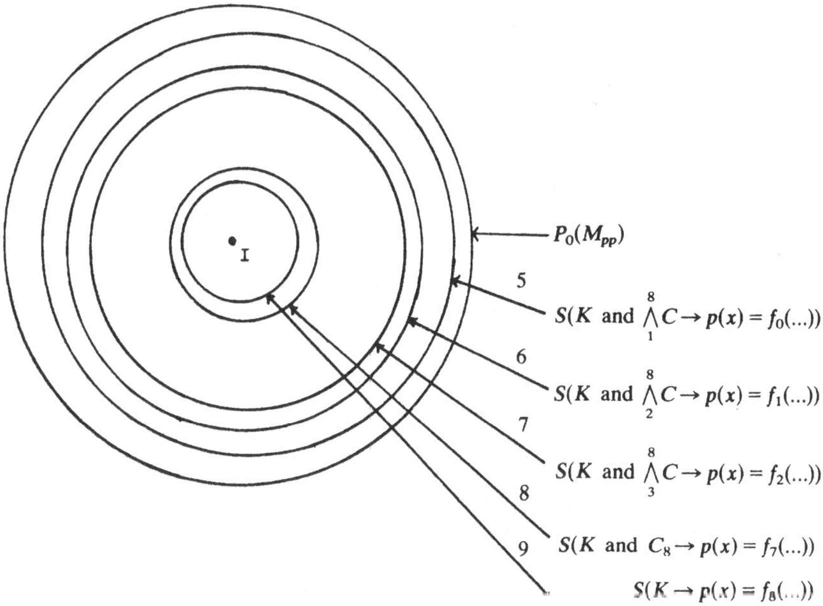

Fig. 2. $P_0(M_{pp})$ (Rearranged).

which coincides with that of the real capitalist economies that occur in history, is in the peel outside even the largest set $S(C_8)$.

And, everybody knows! This is not the kind of claim that Marx sets out to establish.

The Venn-diagram in Figure 2 analyses the sets that are defined by the idealizational statements.

Now, the ideal conditions appear in the antecedent clause of a material implication. And this exactly reverses the relationships between the sets. You simply know that an idealizational statement is true for real capitalist economies because a material implication is true whenever the antecedent clause is false.

So, the fundamental, initial, idealizational statement yielding the set indicated by arrow 5 only rules out an outer peel of which everybody knows that I is not there. If you drop condition 1, you arrive at another set, indicated by arrow 6, and again, everybody is perfectly sure that I is in this set, too, because everybody knows that real economies do not only violate (C_1), they also violate (C_2) to (C_8), the antecedent clause therefore still is false, and the implication therefore still is true for real economies.

There is an inclusion relation between the sets indicated by arrows 5 and 6, because of the reduction relation between f_1 and f_0, and similarly for every f_i and f_{i-1}. So now, all new sets are smaller then the old ones, they are included in the old ones, up to the last set, indicated by arrow 9, where all conditions (C_1) to (C_8) are removed.

Here, I must admit that these functions f can not as such be found in *Capital*, because they presuppose a process of *sequential concretisation*:

$$f_0 \rightarrow f_1 \rightarrow \cdots \rightarrow f_8$$

Another type of concretisation is *isolated concretisation*:

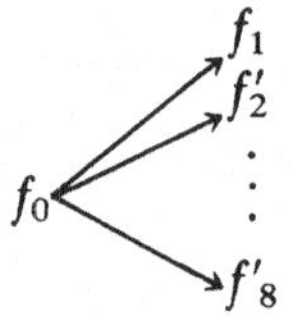

This yields functions f_i' different from f_i because ideal conditions are reintroduced after it is demonstrated how they could be removed. Marx in fact practices a mixed type of concretisation. His actual pathway is thus:

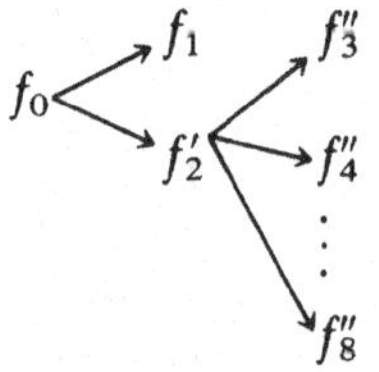

He does so, because in his verbal discourse it would be too cumbersome to retain all deviations caused by, say, merchant capital (C_4) in the treatment of, say, differential ground rent (C_7). But once the values are transformed into production prices (dropping (C_2) which leads to f_2') these production prices have come to stay in the subsequent analysis.

Now what is the harvest of Marx's theory? Not that we now know that the set of real capitalist economies is in the set indicated by arrow

9 of Figure 2. This is not what Marx wished to establish. If the set I would be outside the set indicated by arrow 9 then at least some ideal conditions would have to hold in reality, and we all know they are false for real economies. What does it mean to exclude that I is in the difference of the set indicated by 8 and the set indicated by 9? The final claim:

$$\text{if } K \text{ then } \forall_{x \in E} \, (p(x) = f_8(w(x), \ldots))$$

would be false if there would be capitalist countries where competitive conditions rule and prices do not sell at the eighth sequential modification of values. But this also is uninteresting as a claim of the theory. Competitive conditions hold, says Marx, for capitalist economies only approximately, therefore strictly, *K is false*, and we all know that: there never is completely pure competition on the capital market, the labour market, and the markets for all commodities.

So, Marx is finally prepared to throw away the ladder on which he climbed up from the deep. Also, this turns the eighth modification of the law of value into an untestable statement. Marx considers this final 8th form of the law to be *established*, not by testing but by the procedure of concretisation itself.

A final word about the purpose of modifying the abstract law of value to such a form as is consistent with competitive conditions, that is, to the 8th form. These forms show, says Marx, following Ricardo, that value, though not the *only* determinant of prices, is yet a determinant, and that therefore the *variations* in value determine the *variations* in prices and profits, ceteris paribus. And it is the historical variation of value, that yields the tendency of the profit rate to fall, and allows Marx to predict the end of capitalism.

He does so with the help of trivially true claims that are attained with the help of a complex procedure of concretization. The justification for upholding the ideal law that is the consequent of the claim, is the procedure of concretisation itself.

What can, concluding, be learned from studying Marx, Nowak and Sneed is that

(1) Nowak's series of idealizational statements should not make us forget that Marx has a coherent basic theory from which the ideal conditions are logically deduced.

(2) Though the modified forms of the law of value are not deduced from the basic theory, consistency of these laws with the basic theory is a necessary condition for a successful modification.

(3) Sneed-like claims can be constructed but are trivially true; false antecedent clauses make the law of value and all its modified forms untestable.

(4) The modified forms of the laws are considered – by Marx – to be *established by the procedure of concretisation itself.*

In the final analysis, Marx's book, and for that matter, the book of any other theoretical economist should be looked upon as a chain of proofs. The questions considered by theoretical economists to be the crucial ones are always: what is it that the author wishes to prove, and from what conditions does he prove it, and, finally, what is the relationship between the successive proofs in the chain?

NOTES

* Tilburg University, The Netherlands. The author gratefully acknowledges helpful suggestions of Wolfgang Balzer and other members of the audience of "Philosophy of Economics II".

[1] These conditions slightly differ from those in Nowak (1980), Chap. 1. It is not the purpose of the present paper to treat these differences.

[2] Marx (1894), p. 298.

[3] Nowak gives T_0 to T_8 subscript numbers in reversed order. The order in this article is chosen for convenience; this way, the "i^{th} modification" of the law of value involves function f_i in idealizational statement T_i.

[4] Exactly this metaeconomic requirement is dropped in later attempts to solve the so called transformation problem of values into prices with respect to unequal organic compositions of capital (cf. von Bortkiewicz (1906/7), (1907)).

[5] I will give the exact specifications of the successive f_i's in a forthcoming article in the *Poznan Studies in the Philosophy of the Sciences and the Humanities.*

REFERENCES

Balzer, W.: 1982a, 'Empirical Claims in Exchange Economics' in *Philosophy of Economics*, Springer Verlag, Berlin, W. Stegmüller, W. Balzer and W. Spohn (eds.), pp. 16–40.

Balzer, W.: 1982b, 'A Logical Reconstruction of Pure Exchange Economics', *Erkenntnis* **17**, 23–46.

Balzer, W.: 1985, 'The Proper Reconstruction of Exchange Economics', *Erkenntnis* **23**, 185–200.

v. Bortkiewicz, L.: 1906/7, 'Wertrechnung und Preissrechnung im Marxschen System'. Teil I und II, *Archiv für Sozialwissenschaft und Sozialpolitik*, Band 23, pp. 1–50; Band 25(1), pp. 10–51.

v. Bortkiewicz, L.: 1907, 'Zur Berichtigung der grundlegenden theoretische Konstruktion von Marx im dritten Band des 'Kapital'', *Jahrbücher fur Nationalökonomie und Statistik*, Band 34, pp. 319–55.

Diederich, W. and Fulda, H. F.: 1978, 'Sneed'sche Strukturen in Marx' "Kapital'', *Neue Hefte für Philosophie* **13**, 47–80.

Diederich, W.: 1982, 'A Structuralist Reconstruction of Marx's Economics', in W. Stegmüller, W. Balzer and W. Spohn (eds.), Springer Verlag, Berlin, pp. 145–60.

García de la Sienra, A.: 1982, 'The Basic Core of the Marxian Economic Theory', in W. Stegmüller, W. Balzer and W. Spohn (eds.), Springer Verlag, Berlin, pp. 118–44.

Hands, D. W.: 1985, 'The Logical Reconstruction of Pure Exchange Economics, Another Alternative', *Theory and Decision* **19**, 259–78.

Haslinger, F.: 1982, 'Structure and Problems of Equilibrium and Disequilibrium Theory', in W. Stegmüller, W. Balzer and W. Spohn (eds.), Springer Verlag, Berlin, pp. 63–84.

Haslinger, F.: 1983, 'A Logical Reconstruction of Pure Exchange Economics,' An Alternative View', *Erkenntnis* **20**, 115–129.

Hamminga, B.: 1983, *Neoclassical Theory Structure and Theory Development*, Springer Verlag, Berlin.

Hamminga, B. and Balzer, W.: 1986, 'The Basic Structure of General Equilibrium Theory', *Erkenntnis* **25**, 31–46.

Händler, E. W.: 1980, 'The Logical Structure of Modern Neoclassical Static Microeconomic Equilibrium Theory', *Erkenntnis* **15**, 33–53.

Marx, K.: 1867, *Das Kapital, erster Band*. Berlin, Dietz, 1962.

Marx, K.: 1885, *Das Kapital, zweiter Band*. Berlin, Dietz, 1963.

Marx, K.: 1894, *Das Kapital, dritter Band*. Berlin, Dietz, 1976.

Nowak, L.: 1980, *The Structure of Idealization*. Reidel, Dordrecht etc.

Niiniluoto, I.: 1983, 'Theories, Approximations and Idealizations', *7th LMPS*, Abstracts of Section 6, Salzburg.

Sneed, J. D.: 1976, 'Philosophical Problems of an Empirical Science of Science', *Erkenntnis* **10**, 115–46.

Stegmüller, W., W. Balzer and W. Spohn (eds.): 1982, *Philosophy of Economics*, Proceedings of the Symposium held in Munich, 1981, Springer Verlag, Berlin.

Manuscript received 25 January 1988

Tilburg University
P.O. Box 90153
NL-5000 LE Tilburg
The Netherlands

FURTHER PUBLICATIONS OF THE AUTHORS OF THIS SPECIAL ISSUE

Balzer, W. and J. D. Sneed (1977), 'Generalized Net Structures of Empirical Theories', *Studia Logica* **36** (1977), Part I, *Studia Logica* **37** (1978), Part II.

Balzer, W. (1982), 'Empirical Claims in Exchange Economics', in: Stegmüller et al. (1982), pp. 16–40.

Balzer, W. (1982), 'A Logical Reconstruction of Pure Exchange Economics', *Erkenntnis* **17**, 23–46.

Balzer, W. (1982), *Empirische Theorien: Modelle, Strukturen, Beispiele*, Vieweg Verlag, Braunschweig-Wiesbaden.

Balzer, W., D. Pearce, und H.-J. Schmidt (eds.) (1984), *Reduction in Science*, Reidel, Dordrecht.

Balzer, W. (1985), 'The Proper Reconstruction of Exchange Economics' *Erkenntnis* **23**, 185–200.

Balzer, W. (1985), *Theorie und Messung*, Springer-Verlag, Berlin.

Diederich, W. and H. F. Fulda, (1978), 'Sneed'sche Strukturen in Marx' "Kapital"', *Neue Hefte für Philosophie* **13**.

Diederich, W. (1982), 'A Structuralist Reconstruction of Marx's Economics', in Stegmüller et al. (1982), pp. 115–60.

Garcia de la Sienra, A. (1982) 'The Basic Core of the Marxian Economic Theory' in: Stegmüller et al. (1982), pp. 118–44.

Hamminga, B. (1982), 'Neoclassical Theory Structure and Theory Development: The Ohlin Samuelson Programme in the Theory of International Trade', in Stegmüller et al. (1982), pp. 1–15.

Hamminga, B. (1983), *Neoclassical Theory Structure and Theory Development*, Springer Verlag, Berlin.

Hamminga, B. (1984), 'Possible Approaches to Reduction in Economic Theory' in Balzer et al. (1984), pp. 295–318.

Hamminga, B. and W. Balzer (1986), 'The Basic Structure of General Equilibrium Theory', *Erkenntnis* **25**, 31–46.

Händler, E. W. (1980), 'The Logical Structure of Modern Neoclassical Static Microeconomic Equilibrium Theory', *Erkenntnis* **15**, 33–53.

Händler, E. W. (1980), 'The Role of Utility and of Statistical Concepts in Empirical Economic Theories: the Empirical Claims of the System of Aggregate Supply and Demand Functions Approach', *Erkenntnis* **15**, 129–57.

Händler, E. W. (1982), 'Ramsey Elimination of Utility in Utility Maximizing Regression Approaches', in Stegmüller et al. (1982), pp. 41–62.

Hausman, D. M. (1980), 'How to do Philosophy of Economics', in P. Asquith and Giere (ed.), *PSA 1980.* Philosophy of Science Association, East Lansing, pp. 352–62.

Hausman, D. M. (1981), '*Capital, Profit and Prices: An Essay in the Philosophy of Economics*', Columbia University Press, New York.

Erkenntnis **30** (1989) 267–270.

Hausman, D. M. (1981), 'Are General Equilibrium Theories Explanatory?', in J. Pitt (ed.), *Philosophy in Economics*. Reidel, Dordrecht, pp. 17–32, rpt. in *The Philosophy of Economics: An Anthology*, pp. 344–59.

Hausman, D. M. (1981), 'John Stuart Mill's Philosophy of Economics', *Philosophy of Science* **48**, 363–85.

Hausman, D. M. (1983), 'Are There Causal Relations Among Dependent Variables?', *Philosophy of Science* **50**, 58–81.

Hausman, D. M. (1983), 'The Limits of Economic Science', in N. Rescher (ed.), *The Limits of Lawfulness*. Pittsburgh: Center for Philosophy of Science, University of Pittsburgh, pp. 93–100.

Hausman, D. M. (1984), 'Defending Microeconomic Theory', *Philosophical Forum* **15**, 392–404.

Hausman, D. M. (ed.) (1984) *The Philosophy of Economics: An Anthology*. Cambridge University Press, New York.

Hausman, D. M. (1985), 'Classical Wage Theory and the Causal Complications of Explaining Distribution', in J. Pitt (ed.), *Change and Progress in Modern Science*, *Reidel, Dordrecht*.

Hausman, D. M. (1986), '*Philosophy and Economic Methodology*', *in D. Asquith and P. Kitcher* (*eds.*), *PSA 1984*, vol. 2. East Lansing: Philosophy of Science Association, pp. 231–49.

Janssen, M. (1987), 'Utilistic Reduction of the Macro-economic Consumption Function', *Abstracts LMPS-VIII*, Section II, Moscow.

Kuipers, Th. A. F. (1982), 'Approaching Descriptive and Theoretical Truth', *Erkenntnis* **18**, 343–78.

Kuipers, Th. A. F. (1984), 'Utilistic Reduction in Sociology: The Case of Collective Goods', in Balzer et al. (1984), pp. 239–67.

Kuipers, Th. A. F. (1984), 'Approaching the Truth with the Rule of Success', in P. Weingartner and Chr. Pühringer (eds.), *Philosophy of Science – History of Science*, Selection 7th LMPS Salzburg 1983, *Philosophia Naturalis* **21**, 244–53.

Kuipers, Th. A. F. (1982), 'The Logic of Intentional Explanation', in J. Hintikka and F. Vandamme (eds.), *The Logic of Discourse and the Logic of Scientific Discovery*, Proc. Conf. Gent, 1982, *Communication and Cognition* 18, 177–98.

Kuipers, Th. A. F. (1987), 'A Decomposition Model for Explanation and Reduction', *Abstract LMPS-VIII*, Section 6, Moscow.

Kuipers, Th. A. F. (ed.), (1986), *What is closer-to-the-truth*?, Vol. 10 Bookseries, Poznan Studies in the Philosophy of the Sciences and the Humanities.

Nelson, A. (1984), 'Some Issues Surrounding the Reduction of Microeconomics to Macroeconomics', *Philosophy of Science*, pp. 573–94.

Nelson, A. (1985) 'Physical Properties', *Pacific Philosophical Quarterly*, pp. 268–82.

Nelson, A. (1986), 'Review of *Equilibrium and Macroeconomics*, by Frank Hahn', *Economics and Philosophy*, pp. 148–55.

Nelson, A. (1986), 'Explanation and Justification in Political Philosophy' *Ethics*, pp. 154–76.

Nelson, A. (1986), 'New Individualistic Foundations for Economics', *Nous*, pp. 469–90.

Nelson, A. 'Economic Rationality in Moral Philosophy', *Philosophy and Public Affairs*, forthcoming.

Nowak, L. (1980), 'The Structure of Idealization. Towards a Systematic Interpretation of the Marxian Idea of Science', *Synthese Library*, vol. 139, Reidel, Dordrecht/Boston/London, p. 277.

Nowak, L. (1983), 'Property and Power. Towards a non-Marxian Historical Materialism', *Theory and Decision Library*, vol. 27, Reidel, p. 384.

Nowak, L. (ed.), (1982), *Social Classes, Action and Historical Materialism*, Rodopi, Amsterdam, p. 439.

Nowak, L. and J. Kmita (1970), 'The Assumption of Rationality in Human Sciences', *The Polish Sociological Bulletin*, **1**, 43–68.

Nowak, L. (1973), 'Laws of Science, Theory, Measurement', *Philosophy of Science* **39**, 533–47.

Nowak, L. (1974), 'Value, Idealization, and Valuation', *Quality and Quantity* **7**, 107–19.

Nowak, L. (1977), 'Essence – Idealization – Praxis', *Poznan Studies in the Philosophy of the Sciences and the Humanities* **2**, 1–28.

Nowak, L. (1977), (co-authors P. Chwalisz, P. Kowalik, W. Patryas, and M. Stefanski), 'The Peculiarities of Practical Research', *Poznan Studies in the Philosophy of the Sciences and the Humanities* **2**, 81–100.

Nowak, L. (1977), 'On the Categorial Interpretation of History', *Poznan Studies in the Philosophy of the Sciences and the Humanities* **2**, 1–27.

Nowak, L. (1978), 'Weber's Ideal Types and Marx's Abstractions', in R. Bubner (ed.), 'Marx's Methodologie', *Neue Hefte für Philosophie* 13, Göttingen.

Nowak, L. (1979), 'Idealization and Rationalization: An Analysis of the Anti-Naturalist Programme', in E. Agazzi (ed.), "Specific des sciences humaines en tant que sciences", *Epistemologia*, special issue, vol. 2 pp. 283–305.

Nowak, L. and P. Buczkowski (1980), 'Werte und Gesellschaftklassen', in A. Honneth and U. Jaeggi (eds.), 'Arbeit, Handlung, Normativität', *Theorien des historischen Materialismus*, Bd.2, Suhrkamp, Frankfurt/Main, pp. 365–401.

Nowak, L. (1982), 'On Marxist Social Philosophy', in D. Follesdal (ed.), *Contemporary Philosophy, A New Survey*, vol. 3: 'The Philosophy of Action', Nijhoff, The Hague/Boston/London, pp. 243–75.

Nowak, L. (1985), 'Marxian Historical Materialism: the Case of Dialectical Retardation' in B. Chavance (ed.), 'Marx en perspective', *Editions de l'Ecole des Hautes Etudes en Sciences Sociales*, Paris 1, pp. 77–94.

Nowak, L. (1986), 'Ideology versus Utopia' in P. Buczkowski and A. Klawiter (eds.), 'Theories of Ideology and Ideology of Theories', *Poznan Studies in the Philosophy of the Sciences and the Humanities* **9**, 24–52.

Pearce, D. and M. Tucci (1982), 'On the Logical Structure of Some Value Systems of Classical Economics: Marx and Sraffa', *Theory and Decision* **14**, 155–75.

Pearce, D. and M. Tucci (1982) 'A General Net Structure for Theoretical Economics', in W. Stegmüller, et al. (1982), pp. 85–102.

Pearce, D. and M. Tucci (1984), 'Intertheory Relations in Growth Economics: Sraffa and Wicksell', in W. Balzer, D. Pearce and H.-J. Schmidt (eds.), *Reduction in Science*, D. Reidel, Dordrecht, 1984, pp. 269–93.

Rosenberg, A. (1976), *Microeconomic Laws*, Pittsburg.

Rosenberg, A. (1986), 'Lakatosian Consolations for Economics', *Economics and Philosophy*.

Rosenberg, A. (1986), 'The Explanatory Role of Existence Proofs', *Ethics*.

Rosenberg, A. (1986), 'If Economics isn't Science, What is It?" *Philosophical Forum*.

Sneed, J. D. (1967), 'Strategy and the Logic of Decision', *Synthese* **16**, 392–407.

Sneed, J. D. (1971), *The Logical Structure of Mathematical Physics*, Dordrecht.

Sneed, J. D. and S. Waldhorn (eds.), (1975), *Restructuring the Federal System: Approaches to Accountability in Post-Categorical Programs*, Crane-Russak, New York.

Sneed, J. D. (1975), 'A Defense of Nozick's Method in Anarchy, State and Utopia', *Commentary* p. 20.

Sneed, J. D. (1976), 'John Rawls and the Liberal Theory of Society', *Erkenntnis* **10**, 1–19.

Sneed, J. D. (1978), 'A Utilitarian Framework for Policy Analysis in Food and Food-Related Foreign Aid', in P. G. Brown and H. Shue (eds.), *Food Policy: U.S. Responsibility in the Life and Death Choices*, The Free Press, New York.

Sneed, J. D. (1979), 'Political Institutions as Means to Economic Justice', *Analyse und Kritik* **7**, 125–46.

Sneed, J. D. (1982), 'The Logical Structure of Bayesean Decision Theory', in Stegmüller et al. (eds.) (1982).

Sneed, J. D. and Jamiessen D. (1982), 'What is Quality of Life'? contribution to symposium "Quality of Life in Colorado".

Stegmüller, W., W. Balzer and W. Spohn (eds.), (1982) *Philosophy of Economics*, Proceedings of the Symposium held in Munich, 1981, Springer Verlag, Berlin.

LAKATOS AWARD

The London School of Economics and Political Science has announced that the third Lakatos Award of £10,000 has been awarded to:

Michael Redhead
University of Cambridge
for his book *Incompleteness, Nonlocality & Realism*

The Award, which is for an outstanding contribution to the philosophy of science, has been endowed by the Latsis Foundation in memory of Imre Lakatos. It will be awarded annually for at least ten years.

Professor Redhead will be visiting the L.S.E. to receive his prize and, as a condition of the Award, deliver a public lecture, during the present academic year. The date and the venue will be announced.

Imre Lakatos joined the L.S.E. as a lecturer in 1960, became Professor of Logic in 1969, and died in 1974. He was born in Hungary in 1922. He graduated from Debrecen University in Physics, Mathematics and Philosophy in 1944, whereupon he joined the underground resistance. In 1947 he held the post of Secretary in the Ministry of Education and was virtually in charge of the democratic reform of higher education in Hungary. His prominence got him into trouble: he was arrested in 1950 and spent 3 years in prison. On his release he worked at the Hungarian Academy of Science, where he translated works in the philosophy of mathematics, which introduced him to the subject in which he later became pre-eminent, the logic of mathematical discovery. After the Hungarian uprising he escaped to Vienna and thence to England. He has been described as a 'human dynamo'; one of his most admiring students was Spiro Latsis.

Further information can be obtained from:

Mr Iain Crawford, Press Officer
or Mr Ian Clarke, Press Assistant, L.S.E.
(01-405 7686 Ext. 2053)

CALL FOR MANUSCRIPTS

BRILL'S STUDIES IN EPISTEMOLOGY, PSYCHOLOGY, AND PSYCHIATRY

Brill's Studies in Epistemology, Psychology, and Psychiatry is devoted to the publication of recent philosophical works in these disciplines and, especially, in the areas in which these disciplines intersect. Such works may be of contemporary or historical interest, and of theoretical or practical significance. But they are related in their treatment of philosophical issues and problems pertaining to our understanding of the human mind, its acquisition, validation, and use of knowledge, and the conditions under which such acquisition, validation, and use are or should be regarded as rational.

Should you wish to submit a manuscript for this series, please write to: E. J. Brill, attention Elisabeth Erdman, P.O.B. 9000, 2300 PA Leiden, The Netherlands.

ANNOUNCEMENT

THE ARTHUR PRIOR MEMORIAL CONFERENCE

On 18–23 August 1989 an international joint session of the Australasian Associations of Philosophy and Logic will be held to mark the twentieth anniversary of Prior's death. Papers are invited on Prior, his work, and the subjects on which he wrote (for example – tense logic, modal logic, ethics, determinism, intensionality, ontology, time and change, self reference, indexicals, inferential definition, Aristotelian and medieval logic, logic and religion, epistemic, deontic and erotetic logics). The intention is to publish a selection of the conference proceedings in a Memorial Volume. The venue will be the University of Canterbury, New Zealand, where Prior held the Chair of Philosophy until 1958. Further details may be obtained from Jack Copeland, Philosophy Department, University of Canterbury, Private Bag, Christchurch 1, New Zealand.

Erkenntnis **30** (1989) 273.

ERKENNTNIS

An International Journal of Analytical Philosophy

Erkenntnis is a philosophical journal publishing papers on foundational studies and scientific methodology covering the following areas:

- the field of philosophy associated today with the notions of 'Philosophy of Science' and 'Analytic Philosophy';
- the philosophy of language, of logic, and of mathematics;
- the foundational problems of physics and of other natural sciences;
- the foundations of normative disciplines such as ethics, philosophy of law, and aesthetics;
- the methodology of the social sciences and the humanities;
- the history of scientific method.

Erkenntnis sees one of its objectives as the provision of a suitable platform for the discussion of controversial issues; and another in being a reliable source of timely, competent reviews of important publications in an ever-growing field of research.

Inasmuch as over recent years philosophers standing quite outside the pale of analytic philosophy have also paid careful, and indeed most welcome, attention both to precision of concept and language, and to well-grounded foundations, it is intended that *Erkenntnis* will be a place of meeting, of discussion, and of disputation for philosophers of different persuasions.

Articles for publication, books for review and other communications may be sent to: Professor Wolfgang Spohn, Institut für Philosophie, Universität Regensburg, D-8400 Regensburg, West Germany.

Authors are requested to send manuscripts in threefold.

Erkenntnis is surveyed by Informationsdienst für Philosophie, Revue Philosophique de Louvain, Sociological Abstracts, The Philosopher's Index.

9 780792 301578